Revision and Self Assessme…

ELECTRICAL INSTALLATION

THEORY AND PRACTICE

Part One Studies

MAURICE LEWIS

B Ed (Hons) FIEIE

A member of the Hodder Headline Group
LONDON • SYDNEY • AUCKLAND

First published in Great Britain in 1997 by Arnold,
a member of the Hodder Headline Group,
338 Euston Road, London NWl 3BH.

British Library Cataloguing in Publication Data
A catalogue record for this book is available from the British Library.

ISBN 0 340 67665 5

Typeset in 10/12 Adobe Garamond by GreenGate Publishing Services, Tonbridge, Kent.

Printed and bound in Great Britain by J. W. Arrowsmith Ltd., Bristol

CONTENTS

PREFACE

This self-assessment package is written for the student or trainee pursuing the City & Guilds 2360 Part 1 Electrical Installation Theory and Practice course. It embraces 10 major topics of the current syllabus, beginning with preparation for work and study of the electrical industries and ending with associated electronics technology.

The aim of this assessment package is to assess the student's knowledge and understanding of a wide range of subject matter through short answer multiple-choice questions. These questions are created using the same format as similar questions found in City & Guilds examination papers, i.e. they each have four responses with only one correct answer.

Within each section, as well as a description of topic objectives there are also diagrams and definitions carefully designed to act as precursors to your learning. There are also answers, hints and references at the end of each self assessment.

Before beginning, please read the instruction page on what to do and try to adhere to the generous time limit. This is important and is good practice, because the end-of-course examination will be assessed on a time basis. If you can achieve a percentage pass of 40% in each self-assessment, it is highly likely that you will pass the City & Guilds multiple-choice examination papers.

If you are successful with Assessment 1, try Assessment 2 and Assessment 3. Both are respectively a little harder to do and they will consolidate your learning of the syllabus topic areas.

Finally, considerable information has been given on where to find the answers using various textbooks and other sources. Spend some quiet moments in your college library looking up this information.

Best of luck!

Maurice Lewis

1997

BIOGRAPHICAL NOTE

Maurice Lewis is a well-known author of electrical installation textbooks and currently a City & Guilds Regional Assessor. He has spent many years in the electrical industry and recently retired from teaching where he was Head of the Electrical Section at Luton College of Technology (now Luton University).

Maurice is very much aware of students' difficulty in understanding electrical installation engineering which is forever changing and becoming more and more complex with its technology and regulations. Electricians today have to absorb vast amounts of knowledge about their own discipline; they are not just electrical installers of wiring systems but they have gradually become electrical designers of installations, empowered with much more responsibility for the safety of the electricity consumer.

It is hoped that this self-assessment book will illustrate to the lay person, the amount of knowledge the apprentice must accumulate during the short time spent at college and at work.

1

PREPARATION FOR WORK AND STUDY OF THE ELECTRICAL INDUSTRIES

To tackle the assessments in Topic 1 you will need to know:

- numerous aesthetic effects and environmental controls associated with power stations and transmission lines;
- changes that have occurred in the electricity industry since privatization;
- requirements of some statutory and non-statutory regulations;
- national grid transmission line voltages and voltages at consumer terminals;
- electricity supply and consumer responsibility regarding equipment and switchgear;
- customer requirements during electrical contracting work on their premises;
- standard electrical definitions and safety symbols.

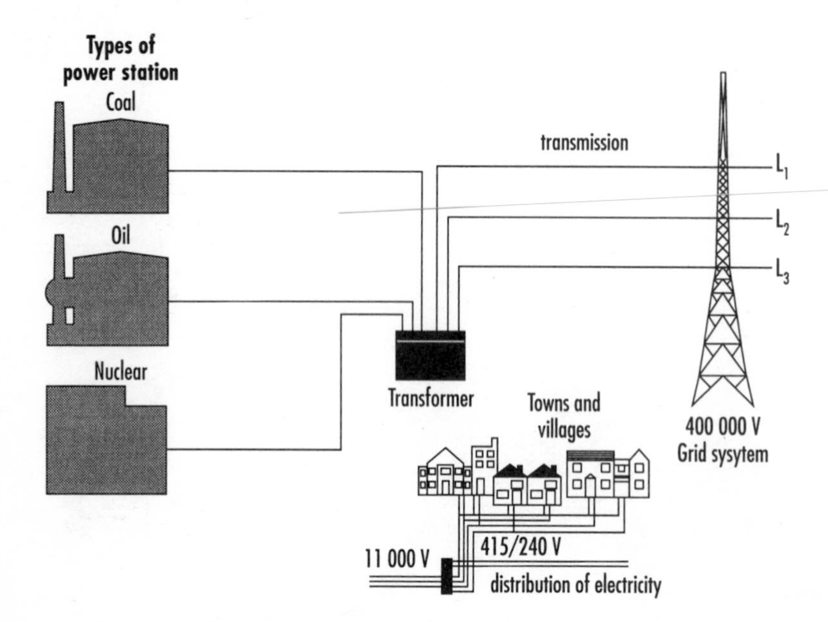

Generation, transmission and distribution of electricity

DEFINITIONS

Accessory – a device other than current-using equipment, associated with such equipment or with the wiring of an electrical installation.
Appliance – an item of current-using equipment other than a luminaire or motor.
Architect – a person employed by a customer/client to carry out, in accordance with contract conditions, certain building works.
Carbon dioxide (CO_2) – a colourless gas with a faint tingling smell and taste which is heavier than air and is emitted from coal-burning power stations.
Clerk of works – acts as the architect's representative on site and ensures that the contractor/subcontractors carry out work in accordance with the drawings and other contract documents.
Consumer unit – a combined fuseboard and main switch controlling and protecting a consumer's final circuit wiring.
Contract – an agreement between two parties which considers and creates an obligation. It is an assurance that work and payment will be carried out in an honest and professional manner.
Distribution board – an assemblage of excess current protective devices in an enclosure with the purpose of protecting final circuits.
Energy meter – an integrating meter that records the number of kilowatt-hours of electricity used by a consumer.

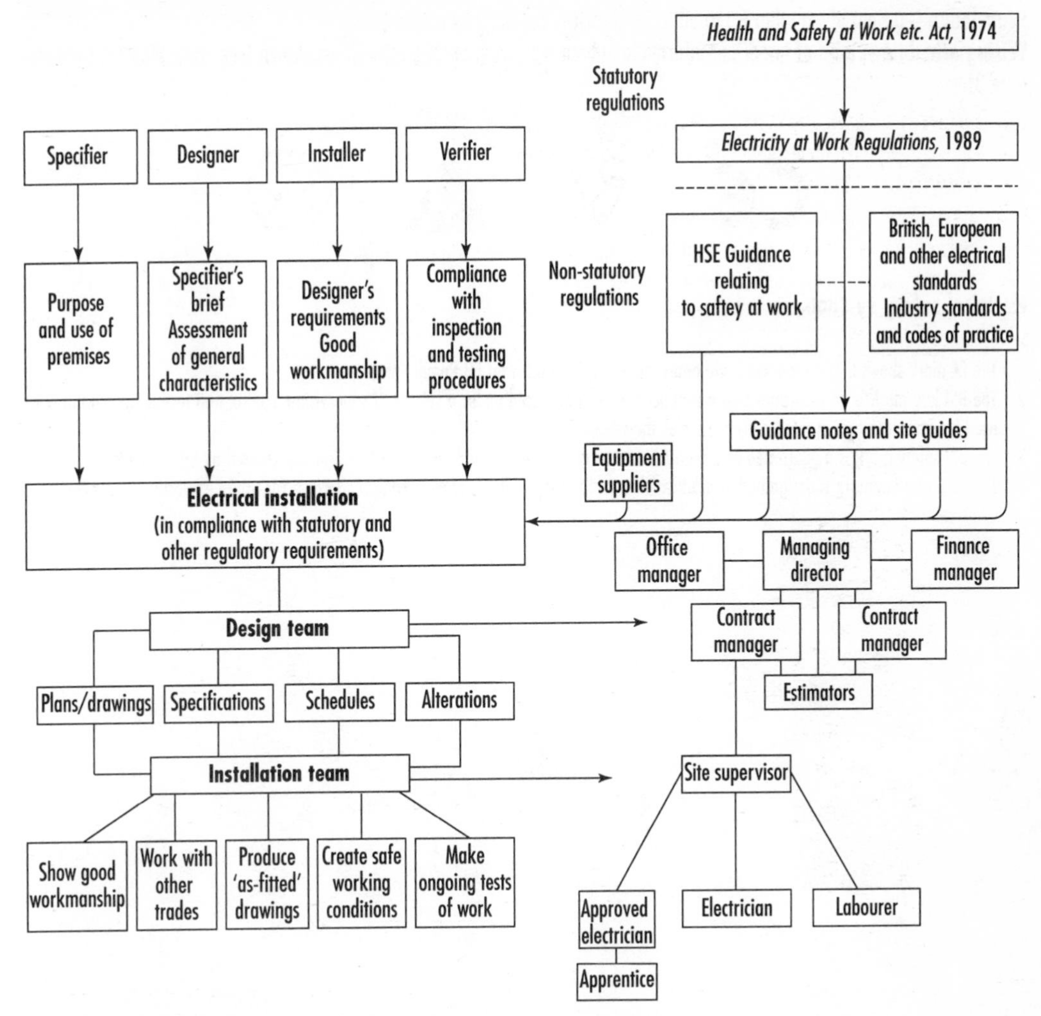

Planning and organisation of electrical installation

Graphite moderator – a component within a nuclear reactor used for slowing down neutrons so that the reactor works more efficiently.
Greenhouse effect – the trapping of reflected infra-red radiation emitted by the Earth, caused by various gases in the atmosphere.
Load centre – populated areas within the country where bulk supplies of electricity are needed.
Luminaire – a lighting fitting that contains all the necessary parts for supporting, fixing and protecting a lamp including controlgear.
National grid – a system of transmission lines and towers used for transporting electricity at high voltage around the country.
Nuclear fission – the bombardment of uranium atoms with slow moving neutron atoms causing a chain reaction and the release of large amounts of nuclear energy.
Reactor – a component within a nuclear power station used for creating heat energy to make steam for driving ac generators.
Rectifier – a piece of electrical apparatus used for converting alternating current electricity into direct current electricity.
Statutory instrument – a regulation passed by Acts of Parliament.
Substation – a building where electricity enters and leaves, often being transformed up or down to different voltage levels.
Sulphur dioxide (SO_2) – a colourless gas which has a choking smell and is emitted from coal-burning power stations.
Transformer – a static piece of electrical equipment mostly used for stepping-up or stepping-down voltage from one level to another.

Way-leave – a right of way that is rented to an electricity company by a land-owner.
Wiring schedule – a list of important information about the wiring of an electrical installation (see Form WR4 IEE Guidance Note 3).

1 2 3 4

Electrical safety symbols

I The CE mark shows that a product is safe and complies with all relevant EU Regulations.
2 The BSI kite-mark is an assurance that a product has been produced under a system of supervision, control and testing and can only be used by manufacturers granted a licence under the scheme.
3 The BSI safety mark is a guarantee of a product's electrical, mechanical and thermal safety and was created by EU Directive.
4 The Ex protection mark is designated to products that conform to standards of certification/approval for use in explosive atmospheres.

TOPIC 1

Preparation for Work and Study of the Electrical Industries

ASSESSMENTS 1.1 – 1.3

Time allowed: 1 hour

Instructions

* You should have the following:

  Question Paper
  Answer Sheet
  HB pencil
  Metric ruler

* Enter your name and date at the top of the Answer Sheet.
* When you have decided a correct response to a question, on the Answer Sheet, draw a straight line across the appropriate letter using your HB pencil and ruler (see example below).
* If you make a mistake with your answer, change the original line into a cross and then repeat the previous instruction. There is only one answer to each question.
* Do not write on any page of the Question Paper.
* Make sure you read each question carefully and try to answer all the questions in the allotted time.

Example:

a 400 V	**a** 400 V
~~**b**~~ 315 V	~~**b**~~ (crossed) 315 V
c 230 V	**c** 230 V
d 110 V	~~**d**~~ 110 V

ASSESSMENT 1.1

1. **Carbon dioxide**, nitrogen oxide and **sulphur dioxide** are all emissions associated with a
   - **a** diesel motorcar engine
   - **b** nuclear-powered submarine
   - **c** supersonic jet aeroplane
   - **d** coal-fired power station.

2. The trapping of the sun's radiation by the atmosphere is called
   - **a** wave diffraction
   - **b** solar eclipse
   - **c** greenhouse effect
   - **d** cause and effect.

3. When an atom of uranium is bombarded with slow moving neutrons (*see* Fig. 1), it breaks up into smaller atoms and starts a process called
   - **a** energy discharge
   - **b** nuclear fission
   - **c** radioactive decay
   - **d** positive charge.

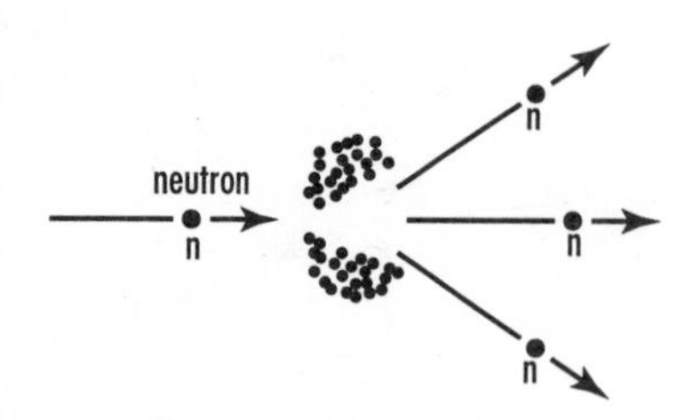

Fig. 1

4. A **way-leave** is a right of way
   - **a** rented out to an electricity company
   - **b** between a wall and an electrical switch-board
   - **c** leading towards an electricity substation
   - **d** illuminated during the hours of darkness.

5. Before privatization of the electricity supply industry, Regional Electricity Companies (RECs) were called
   - **a** Area Boards
   - **b** Generating Boards
   - **c** District Undertakers
   - **d** Supply Distributors.

6. All of the following are transmission line voltages *except*
   - **a** 400 kV
   - **b** 275 kV
   - **c** 132 kV
   - **d** 66 kV.

7. Which of the following is called a 'green-house' gas?
   - **a** nitrogen
   - **b** hydrogen
   - **c** helium
   - **d** carbon dioxide.

8. The system used for transporting electricity from one part of the country to the other is called the

a supply route
b super highway
c national grid
d network link.

9. Power stations are often sited near coastal locations to provide an

a unobtrusive effect on the countryside
b abundance of cooling sea water
c underwater electricity link to Europe
d efficient method of heating seawater.

10. The equipment shown in Fig. 2 which is used to change ac voltage from one level to another is called a

a transformer
b rectifier
c regulator
d converter.

Fig. 2

11. A building that contains high and low voltage electrical equipment for the sole purpose of changing voltage levels is called a

a load centre
b relay room
c substation
d power house.

12. All of the following are standard low voltage ac supplies found in use in the UK, *except*

a 415 V
b 350 V
c 240 V
d 110 V.

13. All of the following are advantages of using electricity over other utility services *except*

a artificial lighting
b information/data storing
c labour-saving appliances
d heating for comfort.

14. Which one of the following devices is *not* an electrical **appliance**?

a freezer motor
b hair dryer
c food mixer
d vacuum cleaner.

15. A **luminaire** is defined as a

a solar energy panel
b lighting fitting
c sowing machine
d radiant heater.

16. Which symbol shown in Fig. 3 shows that a product is safe and that it complies with all relevant European Union Regulations?

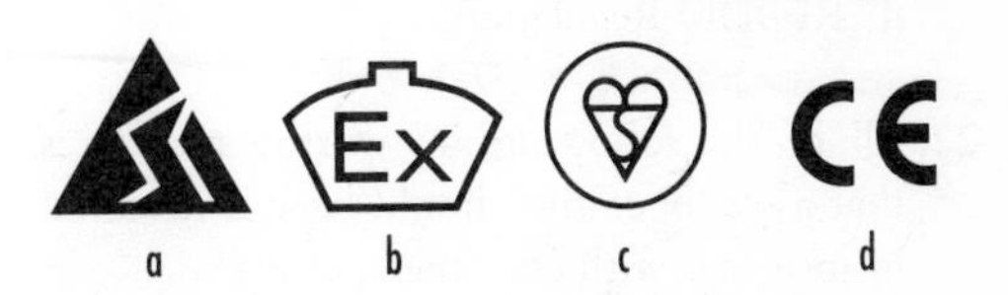

Fig. 3

17. The **statutory instrument** dealing with electrical equipment and all work situations where electricity is used is called the

a IEE Wiring Regulations
b Electricity at Work Regulations
c Health and Safety at Work Act
d Electricity Supply Regulations.

18. The professional person who acts on behalf of the customer in a building project is called the

a clerk of works
b architect
c quantity surveyor
d site manager.

19. An electrical apprentice in his third year of training is called

a an indentured apprentice
b a competent apprentice – Stage 1
c a senior apprentice – Stage 2
d a journeyman apprentice.

20. Which one of the following is the *least* important when trying to maintain good working relations on a construction site?

a good time keeping and behaviour
b knowledge of safety regulations
c knowledge of site deliveries
d respecting the work of others.

21. An employee has a statutory duty on site 'not to intentionally or recklessly interfere with or misuse anything provided in the interest of health and safety'. This is a requirement of the

a IEE Regulations
b HSW Act
c COSHH Regulations
d RIDDO Regulations.

22. All of the following are worthy attributes that a site operative should have if involved in meetings with customers, *except*

a understanding of contractual procedures
b acting conscientiously and being polite
c being properly dressed wearing suitable overalls
d knowing the names and positions of the company's managers.

23. Which one of the following is *not* a factor to be borne in mind when completing a customer's electrical rewiring?

a leaving behind instructional information
b leaving the premises inspected and tested
c restoring the property at the end of the work
d making sure the kWh energy meter is connected.

24. Fig. 4 shows the sequence of control equipment at the intake position of a consumer's premises. The cut-out fuse (1) and energy meter (2) are the responsibility of the

a electrical contractor
b electricity user
c building insurance company
d regional electricity company.

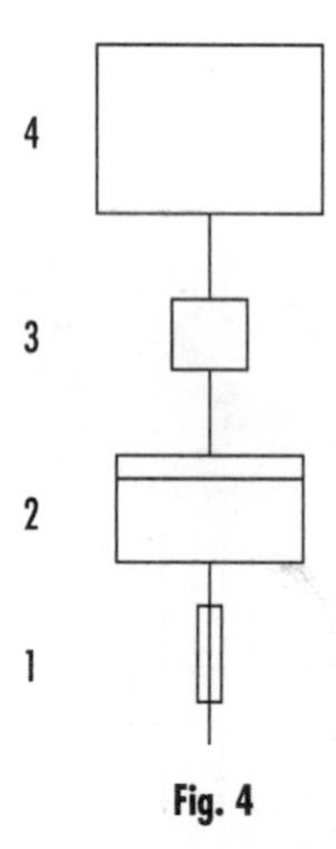

Fig. 4

25. Name the document that provides an assurance by both contractor and customer that work and payment will be carried out in an honest and professional manner

a permit to work
b formal contract
c approval notice
d schedule of daywork.

ASSESSMENT 1.1

Answers, Hints and References

1. **d** *See* definition and Ref. 1, pp 4–8.
2. **c** *See* definition and Ref. 1, pp 4–5.
3. **b** *See* definition.
4. **a** *See* definition and Ref. 1, p 7.
5. **a** *See* Ref. 1, p 1.
6. **d** It is a non-standard voltage.
7. **d** *See* definition and Ref. 1, pp 4–5.
8. **c** *See* definition and Ref. 1, p 1.
9. **b** *See* Ref. 1, p 7.
10. **a** *See* definition: **b** changes ac into dc by static means; **c** keeps voltage constant; and **d** changes ac to dc by means of a motor generator set.
11. **c** *See* definition: **a** is a suitable place to take bulk supplies of electricity from the Grid; **b** is a room containing monitoring instruments; and **d** is a room containing electrical switchgear.
12. **b** Note words in question: 'in use'.
13. **d** Gas appliances and oil appliances are all available.
14. **a** *See* definition and note that it is an independent electrical item.
15. **b** *See* definition and Ref. 1, p 126.
16. **d** *See* definition. It is a new symbol.
17. **b** Note that the IEE Wiring Regulations are non-statutory.
18. **b** *See* definition and Ref. 2, p 95.
19. **c** *See* Ref. 1, p 131, Fig. A5.1.
20. **c** An extraneous matter outside an individual's control.
21. **b** *See* Ref. 1, p 13.
22. **a** It is not part of his education and training.
23. **d** It is outside an installer's remit.
24. **d** *See* Ref. 1, p 40.
25. **b** *See* definition and Ref. 2, p 15.

ASSESSMENT 1.2

1. Which one of the following power stations emits vast quantities of **sulphur dioxide** into the atmosphere?

   a nuclear power station
   b gas-fired power station
   c coal-fired power station
   d hydro-electric power station.

2. In the gas-cooled reactor shown in Fig. 1, the chain reaction is controlled by movable control rods made from a material called

   a boron
   b iron
   c carbon
   d silicon.

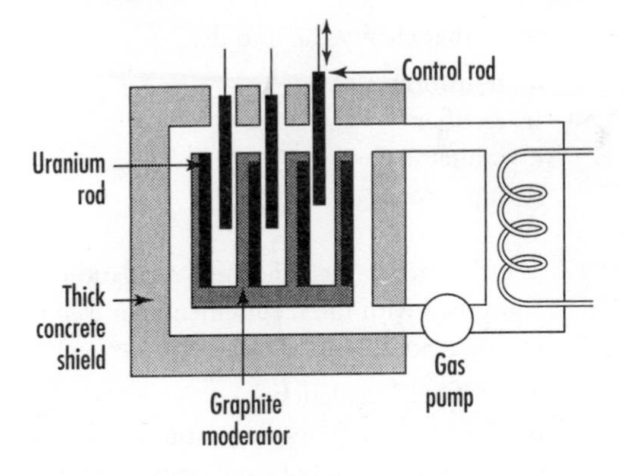

Fig. 1

3. The purpose of the **graphite moderator** in Fig. 1 is to

   a absorb uranium atoms
   b slow down neutrons
   c heat the reactor core
   d protect the fuel rods.

4. A piece of land rented to an electricity company by a landowner to install transmission lines is called a

   a footpath
   b way-leave
   c walkway
   d tow-path.

5. MANWEB, NORWEB, SWALEC and SEEBOARD are abbreviations of four

   a regional electricity companies
   b electricity generating boards
   c electricity supply authorities
   d area electricity suppliers.

6. What is the correct name given to the tall structure shown in Fig. 2?

   a electricity pillar
   b national grid mast
   c high voltage column
   d transmission tower.

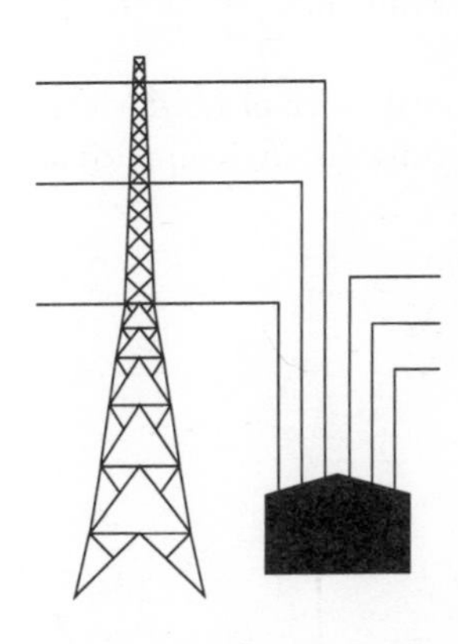

Fig. 2

7. The national grid network in Britain operates mostly at a voltage of

  a 400×10^3 V
  b 400×10^6 V
  c 400×10^9 V
  d 400×10^{12} V.

8. The term used on the **national grid** system that describes the bulk supply of electricity to consumers is a

  a substation
  b demand point
  c load centre
  d take-off point.

9. Which of the following poses the *least* problem for choosing a site for a power station?

  a insufficient acreage of land
  b obtrusion on the countryside
  c short distance to the coast
  d accessibility to roads and railways.

10. Voltage changes created by electromagnetic induction are from equipment called

  a transformers
  b rectifiers
  c regulators
  d converters.

11. A room that contains high and low voltage electrical equipment for the sole purpose of changing voltage levels is called a

  a transformer room
  b relay room
  c switch room
  d power room.

12. A nominal, three-phase ac voltage to industrial premises is often supplied at

  a 615 V
  b 500 V
  c 415 V
  d 380 V.

13. The symbol shown in Fig. 3 identifies electrical equipment that is

  a double insulated
  b internally earthed
  c made in non-EC countries
  d supplied at low voltage.

Fig. 3

14. Which one of the following devices is *not* an electrical accessory?

  a conduit bush
  b cable saddle
  c hair dryer
  d adaptable box.

15. What is the name given to a light fitting and its lamp?

  a ballast
  b choke
  c ignitor
  d luminaire.

16. What is the name given to electrical equipment that changes ac into dc?

  a transformer
  b rectifier
  c inductor
  d capacitor.

17. Which one of the following regulations is concerned with the requirements for electrical installations?

  a COSSH Regulations
  b Electricity at Work Regulations
  c Electricity Supply Regulations
  d IEE Wiring Regulations.

18. An electrical installation design team is responsible for all of the following, *except*

a plans and drawings
b good workmanship and on-going tests
c specifications and schedules
d alterations and deviations.

19. The person acting on behalf of an architect on a large building site is called a

a chargehand
b clerk of works
c general foreman
d work's engineer.

20. All of the following should be avoided when working on a construction site, *except*

a taking short cuts to speed up work
b only thinking of one's own safety
c listening to a personal stereo system
d wearing a hard hat at all times.

21. Electrical contracting work carried out on a customer's premises should always be to

a create minimum disruption to normal activities
b provide high quality work at the expense of profit
c provide a permanent electricity supply at all cost
d achieve further work at a later stage.

22. Fig. 4 shows the control equipment at the intake position. The circuit breaker (3) and distribution board (4) are the overall responsibility of the

a electrical installation contractor
b consumer or owner of the building
c building insurance company
d regional electricity company.

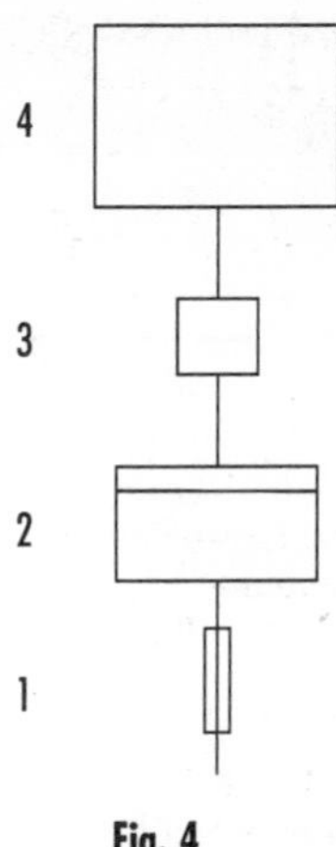

Fig. 4

23. When an electrical installation is being wired, the installer is responsible for all the following connections *except* the

a earth electrode connection
b main switch connection
c earth bonding connection
d cut-out connection.

24. A formal building contract is an agreement between parties that

a recognizes a shared commitment to work
b contemplates and creates an obligation
c is granted only by local authority approval
d identifies all work to be done in a project.

25. Which of the following is *not* classified as being a contract document?

a bills of quantities
b plans and drawings
c schedules
d wiring regulations.

ASSESSMENT 1.2

Answers, Hints and References

1. **c** *See* definition and Ref. 1, pp 4–8.
2. **a** Boron is a brown amorphous (non-crystalline) powder and has the property of absorbing slow neutrons.
3. **b** *See* definition.
4. **b** *See* definition.
5. **a** *See* Ref. 3, p 13.
6. **d** *See* Ref. 1, p 3, Fig. 1.2.
7. **a** *See* Ref. 1, p 1.
8. **c** *See* definition and Ref. 1, p 1.
9. **c** *See* Ref. 1, p 7.
10. **a** *See* definition: **b** changes ac into dc by static means; **c** keeps voltage constant; and **d** changes ac to dc by means of a motor generator set.
11. **a** *See* definition.
12. **c** *See* Ref. 4, voltage definitions.
13. **a** *See* Ref. 1, p 114.
14. **c** *See* definition.
15. **d** *See* definition.
16. **b** *See* definition.
17. **d** Note that the IEE Wiring Regulations are non-statutory.
18. **b** See Figs. T1 and T2, pp 2–3.
19. **b** *See* definition.
20. **d** Wearing a hard hat is often compulsory on a construction site.
21. **a**.
22. **b**.
23. **d** It is outside an installer's remit.
24. **b** *See* definition and Ref. 2, p 15.
25. **d** *See* Ref. 2, p 15.

ASSESSMENT 1.3

1. Which one of the following is *not* a chimney emission from a large coal-fired power station?

   a nitrogen oxide
   b carbon dioxide
   c lead peroxide
   d sulphur dioxide.

2. Radioactive waste is often associated with

   a an oil power station
   b a coal power station
   c a gas power station
   d a nuclear power station.

3. The term **way-leave** describes

   a a footpath running across agricultural land for use by the general public
   b a parcel of land rented to an electricity company by a landowner
   c a thoroughfare leading towards an electricity substation
   d an illuminated tow-path running at the side of a canal.

4. Under privatization of the electricity industry the Central Electricity Generating Board (CEGB) was taken over by the

   a National Grid Company
   b Area Electricity Boards
   c Nuclear Electric plc
   d Regional Electricity Companies.

5. In Fig. 1 the building adjacent to the transmission tower is called a

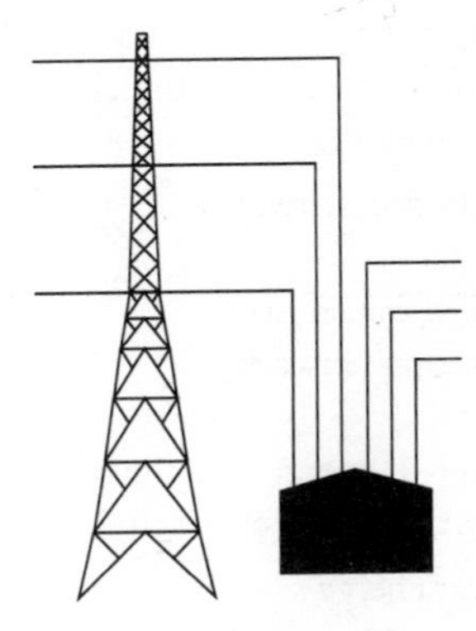

Fig. 1

   a relay chamber
   b generating room
   c power house
   d substation.

6. The reason why direct current electricity is *not* transmitted by the national grid system is because

   a lines and equipment would be too large
   b supply voltages need to be transformed
   c earth faults would go undetected
   d consumers need high frequency supplies.

7. Which group of ac transmission voltages (in kV) is commonly found in Britain?

   a 400, 275, 132, 33
   b 420, 275, 133, 66
   c 415, 240, 132, 11
   d 400, 350, 250, 6.6.

8. Which group of ac low voltages (in kV) is often used by electricity consumers?

- **a** 315, 230, 120, 25
- **b** 500, 220, 110, 50
- **c** 415, 240, 110, 64
- **d** 400, 240, 64, 50.

9. Which one of the following is *not* regarded as a greenhouse gas?

- **a** carbon dioxide
- **b** hydrogen
- **c** methane
- **d** nitrous oxide.

10. The system that is used for transporting electricity from one part of the country to the other is called the

- **a** distribution network
- **b** national grid
- **c** regional transmission
- **d** generating network.

11. What type of energy uses the sun's radiation to convert heat into electricity?

- **a** nuclear energy
- **b** solar energy
- **c** geothermal energy
- **d** kinetic energy.

12. When ac electricity is transformed from a low voltage level to a high voltage level, the current that is transformed

- **a** falls to a low level
- **b** remains constant
- **c** increases to a high level
- **d** becomes unidirectional.

13. In a power station the room that is used to monitor the flow of electricity is called a

- **a** main room
- **b** relay room
- **c** control room
- **d** switch room.

14. The marketplace for trading electricity that allows for competition in generation and supply is called a

- **a** pool
- **b** grid
- **c** centre
- **d** nest.

15. The symbol shown in Fig. 2 denotes electrical equipment that is safe and

- **a** continuously earthed
- **b** required by EU Regulations
- **c** double insulated
- **d** supplied at low voltage.

Fig. 2

16. Food mixers, hair dryers, vacuum cleaners etc. are all examples of electrical

- **a** accessories
- **b** apparatus
- **c** appliances
- **d** equipment.

17. Supply plant that is used to keep consumers' voltage steady is called a

- **a** full-wave rectifier
- **b** tap-change transformer
- **c** synchronous machine
- **d** static converter

18. Which one of the following regulations deals with requirements for electricity supplies?

- **a** COSSH Regulations
- **b** Electricity at Work Regulations
- **c** Electricity Supply Regulations
- **d** IEE Wiring Regulations.

19. The symbol shown in Fig. 3 is a

- **a** BASEEFA mark for explosive equipment
- **b** ROSPA safety mark for unsafe floors
- **c** CENELEC approval mark for test equipment
- **d** BSI safety mark for electric appliances.

Fig. 3

20. All of the following are advantages of using electricity over other utilities such as gas, water and oil *except*

a artificial lighting
b information/data storing
c labour-saving appliances
d heating for comfort.

21. The person responsible for producing 'as fitted' drawings on a construction site is a

a site agent
b site foreman
c site safety officer
d site office clerk.

22. Electrical contracting work carried out on a customer's premises should always be to

a create minimum disruption to normal activities
b provide high quality work at the expense of profit
c maintain a permanent electricity supply at all cost
d achieve further work at a later stage.

23. When electrical wiring is completed on a customer's premises it is important to

a leave behind the bill for carrying out the work
b leave your address for carrying out further work
c restore the property on completion of the work
d make sure all electrical circuits have been inspected and tested.

24. A formal agreement between parties that contemplates and creates an obligation is called

a a specification
b a contract
c a schedule of work
d an estimate.

25. In the process of re-wiring a furnished house, you accidentally break an expensive glass vase. Which one of the following should you do first?

a inform nobody about the matter
b inform your employer after work finishes
c inform the householder if present
d inform your electrician immediately.

ASSESSMENT 1.3

Answers, Hints and References

1. **c** Often found as an essential substance in a lead-acid cell, see Ref. 1, pp 4–8.
2. **d** *See* Ref. 1, p 1.
3. **b** *See* Ref. 1, p 6.
4. **a** *See* Ref. 1, p 1.
5. **d** *See* Ref. 1, p 3, Fig. 1.2.
6. **a** *See* Ref. 1, p 1.
7. **a** *See* Ref. 1, p 7.
8. **c** *See* Ref. 1, p 37.
9. **b** *See* Ref. 1, p 7.
10. **b** *See* Ref. 1, p 1.
11. **b** *See* Ref. 1, p 7.
12. **a** *See* Ref. 6, p 45.
13. **c**.
14. **a** *See* Ref. 5.
15. **b** It is a new symbol.
16. **c** *See* stated definition.
17. **b**.
18. **c** *See* Ref. 1, p 36.
19. **d** It is seen on electrical appliances.
20. **d** The other energy sources can also be used for this provision.
21. **b** He is the person in charge on site.
22. **a** It is wholly in the interest of the customer.
23. **d** It is outside an installer's remit.
24. **b** *See* Ref. 2, p 15.
25. **d** He is your immediate supervisor.

2

ASSOCIATED CORE SCIENCE

To tackle the assessments in Topic 2 you will need to know:

- definitions of quantities and units relating to electrical science and mechanical science;
- different types of force, principles of moments, equilibrium and classes of lever;
- basic formulae used to solve electrical and mechanical problems;
- principles of thermometers and how to solve problems relating to temperature changes;
- effects of electric current and how to solve circuit problems relating to Ohm's law;
- ac circuit theory relating to the property of sinusoidal waveforms;
- standard electrical circuit symbols.

DEFINITIONS

Capacitor – a component consisting of two metal plates separated by an insulating layer called a dielectric and having the ability to store an electric charge.
Commutator – an assembly of conducting members insulated from each other against which brushes bear, used to enable current to flow from one part of a circuit to another by sliding contact.
Coulomb – the quantity of electricity crossing a section of conductor in a time of one second (6.3×10^{18} electrons).
Current – the flow or transport of negative charges (electrons) along a path or around a circuit, measured in amperes.
Density – the mass per unit volume of a body or substance measured in kilograms per cubic metre.
Diode – a semiconductor pn junction rectifier.

Mensuration	Electrical	Mechanical
$A = \pi r^2$	$Q = It$	$F = mg$
$C = \pi d$	$R = V/I$	$F = BIL$
$V = Ah$	$R = \rho L/A$	$F_1 = F_2 D_2/D_1$
$a^2 = b^2 + c^2$	$P = I^2R$	$C = 5/9(F - 32)$
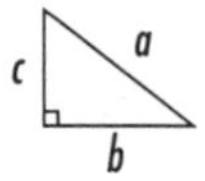	$f = 1/T$	$F = 9/5C + 32$
	$n = f/p$	$\rho = m/V$

Common formulae

Eddy currents – are induced circulating currents set up in the metal components of transformers and machines causing unnecessary heating and power losses.
Effective value – the root mean square value of an ac sine wave, being 0.707 times the maximum or peak value; causes the same heating effect in a resistor as direct current.
Efficiency – the ratio of work output to work input expressed in the same units and usually stated as a percentage or per unit value.
Electromotive force – the force necessary to cause the movement of charges.
Energy – the capacity for doing work such as potential energy (e.g. coiled spring) or kinetic energy (e.g. the spring being released) measured in joules.
Force – the external agency capable of altering the state of rest in a body through a 'push' or 'pull' action, measured in kilograms.
Frequency – the number of repetitive cycles generated in one second, measured in hertz.
Fulcrum – the point of support on which a lever pivots.
Inductance – the property of an inductor producing a magnetic field when carrying current, measured in henries.
Magnetic flux – the phenomenon associated with invisible lines of force in the neighbourhood of magnets and electric currents, measured in webers.

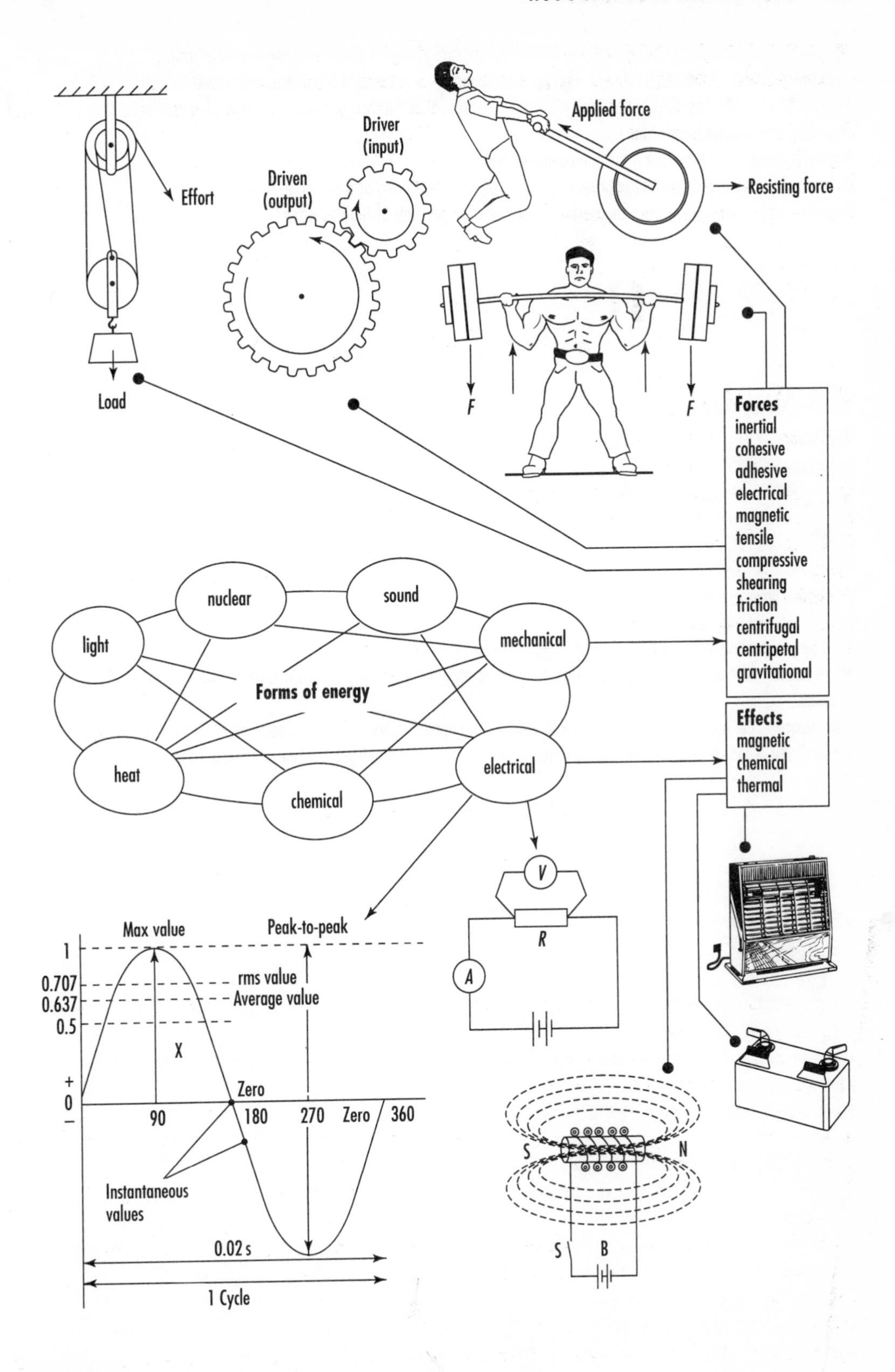

Forms of energy

Magnetic flux density – the quantity of magnetic flux spread over a given area, measured in teslas.
Maximum value – the peak value or highest instantaneous value reached by a generated waveform.
Ohm's Law – the law that states that the ratio of potential difference between the ends of a conductor and the current flowing in a conductor is constant.
Periodic time – the time taken for one complete cycle to occur.
Potential difference – a voltage pressure drop across different parts of a circuit, measured in volts.
Power – the rate of doing work measured in units of work per unit of time or watts.

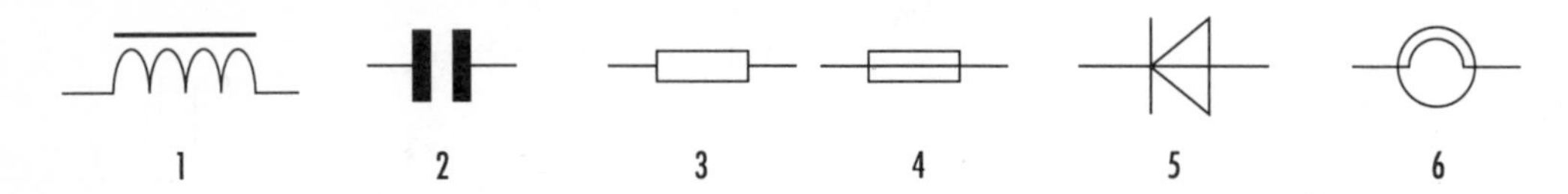

Electrical wiring diagram symbols

1 Inductor

2 Capacitor

3 Resistor

4 Fuse

5 Rectifier

6 Filament lamp

Proton – a positively charged particle residing in the nucleus of an atom.
Resistance – the opposition to current flow in resistive components, expressed as a ratio of V/I, (see Ohm's Law) measured in ohms.
Reactance – the opposition to current flow in inductive components, expressed as a ratio of V/I, measured in ohms.
Resistivity – the resistance between the opposite faces of a unit cube of given material measured in ohm metres.

TOPIC 2

Associated Core Science

ASSESSMENTS 2.1 – 2.3

Time allowed: 1½ hours

Instructions

* You should have the following:

  Question Paper
  Answer Sheet
  HB pencil
  Metric ruler

* Enter your name and date at the top of the Answer Sheet.
* When you have decided a correct response to a question, on the Answer Sheet, draw a straight line across the appropriate letter using your HB pencil and ruler (see example below).
* If you make a mistake with your answer, change the original line into a cross and then repeat the previous instruction. There is only one answer to each question.
* Do not write on any page of the Question Paper.
* Make sure you read each question carefully and try to answer all the questions in the allotted time.

Example:

a	400 V	**a**	400 V
~~**b**~~	315 V	~~**b**~~ (crossed)	315 V
c	230 V	**c**	230 V
d	110 V	~~**d**~~	110 V

ASSESSMENT 2.1

1. All of the following are SI units, *except*

a volume
b mass
c length
d time.

2. Inertial force is the force needed to

a hold things together
b stretch things apart
c stop and start things
d make things act downwards.

3. In Fig. 1, the mass rests on the surface owing to the Earth's gravitational force of

a $9.81\ m/s^4$
b $9.81\ m/s^3$
c $9.81\ m/s^2$
d $9.81\ m/s^1$.

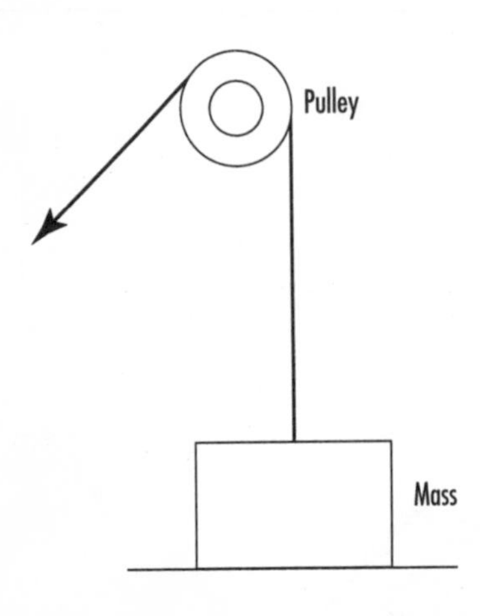

Fig. 1

4. The kilogram is the unit of

a density
b mass
c weight
d volume.

5. What is the circumference of a copper bus-bar measuring 50 mm in diameter?

a 157.1 cm
b 15.71 cm
c 1.571 cm
d 0.157 cm.

6. In Fig. 2, what is the resultant force (F_R)?

a 8 N
b 7 N
c 6 N
d 5 N.

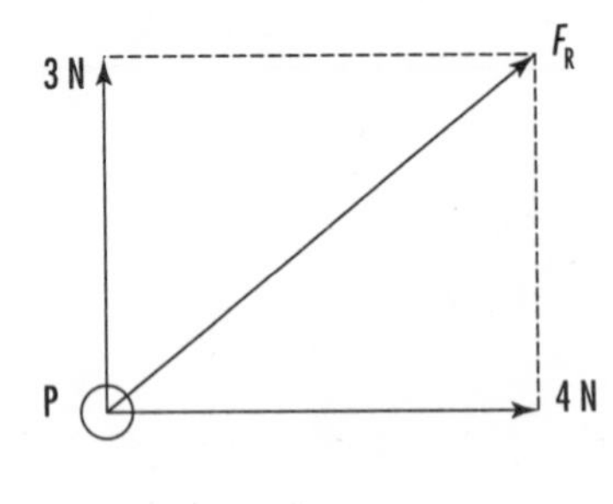

Fig. 2

7. What is the derived SI unit for electric charge?

a ampere
b volt
c farad
d coulomb.

8. Fig. 3 illustrates the principle of moments. To create a balance

a $F_1 + D_1 = F_2 + D_2$
b $F_1 \times D_1 = F_2 \times D_2$
c $F_1 - D_1 = F_2 - D_2$
d $F_1 \div D_1 = F_2 \div D_2$.

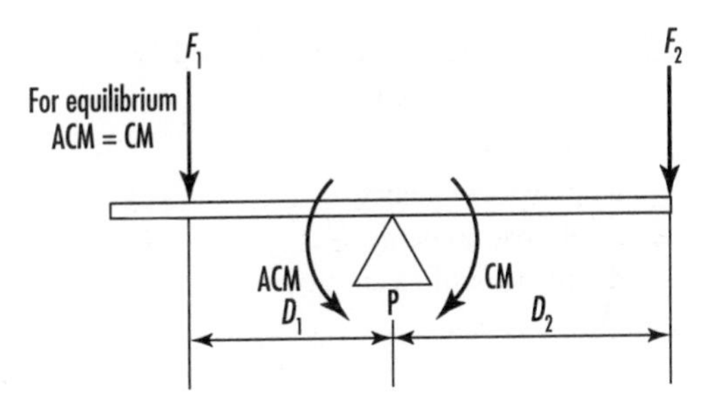

Fig. 3

9. The force per unit area acting on a surface and measured in newtons per square metre is called

a stress
b compression
c pressure
d friction.

10. A **proton** is a

a negatively charged particle
b positively charged particle
c neutral particle
d ionized particle.

11. Mercury is used in a thermometer because it expands uniformly and also because it

a is a non-toxic substance
b is a good conductor of heat
c boils at 200°C
d freezes at –200°C.

12. Which one of the following is the freezing point of water in degrees Celsius?

a 32°C
b 10°C
c 0°C
d –5°C.

13. A **coulomb** of electricity is equivalent to

a 11×10^3 volts
b 6.3×10^{18} electrons
c 3.6×10^6 joules
d 1.5×10^{-3} amperes.

14. Heat transference through solids is called

a conduction
b generation
c convection
d radiation.

15. Which one of the following formulae is an expression for **Ohm's Law**?

a $R = V/X$
b $R = V/Z$
c $R = V/I$
d $R = V/S$.

16. In Fig. 4, the potential difference across R_2 can be expressed as

a $V_2 = V_s + V_1 + V_3$
b $V_2 = V_s + V_1 - V_3$
c $V_2 = V_s - V_1 - V_3$
d $V_2 = V_s - V_1 + V_3$.

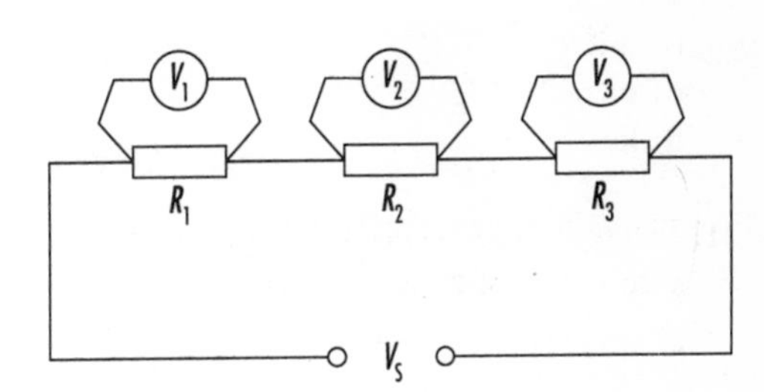

Fig. 4

17. The current taken by a 1 kW, 230 V halogen lamp can be found using the formula

a $I = E/P$
b $I = E/Z$
c $I = P/V$
d $I = W/V$.

18. All the following are factors that influence a conductor's resistance, *except*

a length
b resistivity
c area
d impedance.

19. In Fig. 5, if switches S1 to S4 are open and switches S5 to S7 closed, the circuit resistance would be

a 60 Ω
b 48 Ω
c 8 Ω
d 6 Ω.

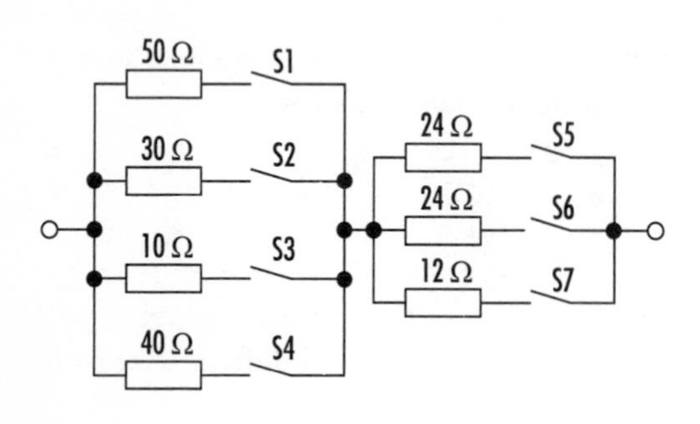

Fig. 5

20. The prefix name for the number 10^{-9} is

a nano
b micro
c pico
d hecto.

21. The resistance between the opposite faces of a unit cube of material is called

a impedance
b density
c resistivity
d conductance.

22. A wattmeter is used in a circuit to measure

a energy used
b consumed power
c supply frequency
d conductor resistance.

23. The electricity produced by a battery or accumulator is known as

a direct current
b alternating current
c single-phase current
d positive-negative current.

24. A supply generated at a frequency of 50 Hz, passes through zero one hundred times every

a day
b hour
c minute
d second.

25. In Fig. 6, the instrument marked W measures the circuit's

a power factor
b power consumed
c electrical energy
d quantity of electricity.

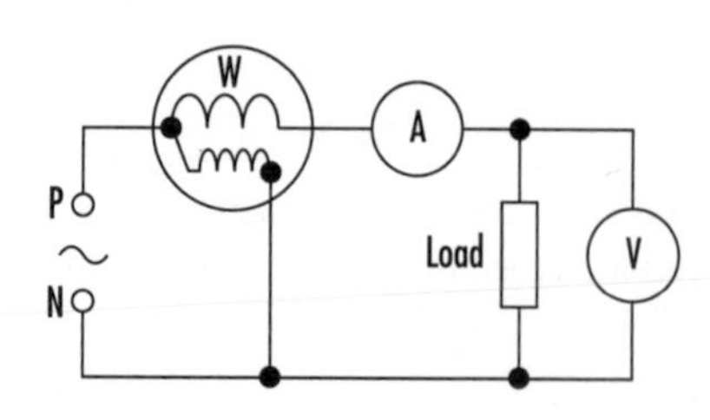

Fig. 6

26. In a right-angled triangle, the square on the hypotenuse is equal to the sum of the squares on the other two sides. This is known as

a Kirchhoff's law
b Archimedes' principle
c Pythagoras' theorem
d Charles' law.

27. The derived SI unit for **magnetic flux density** is called the

a weber
b tesla
c henry
d siemen.

28. When two permanent magnets are closely positioned such that their opposite poles face each other, the effect is for them both to

a repel
b stand on end
c attract
d move at 30°.

29. In Fig. 7, the formula used to find the **force** on the conductor is

a $F = mg$
b $F = BIL$
c $F = BLv$
d $F = W/L$.

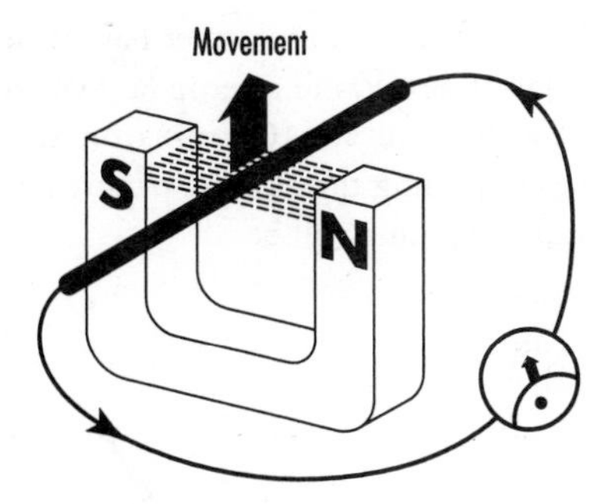

Fig. 7

30. When two closely parallel conductors pass current in opposite directions the effect will be to

a attract each other
b repel each other
c rotate around each other
d remain stationary.

31. Fig. 8 shows a sinusoidal voltage quantity taken over one complete cycle (1 Hz) of the supply. The number of cycles completed in one second is called the

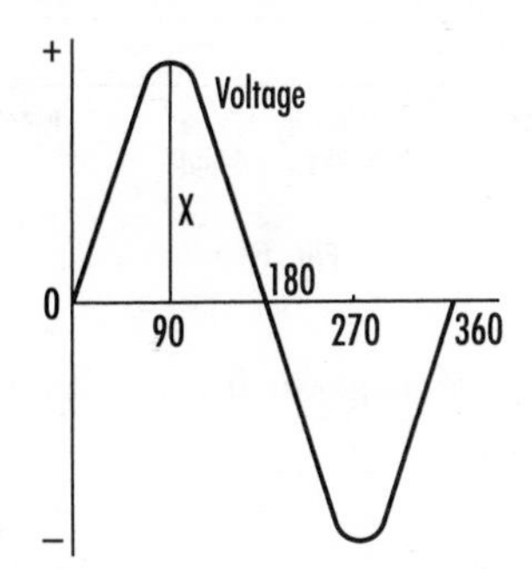

Fig. 8

a period
b waveform
c frequency
d amplitude.

32. In Fig. 8 the line marked X is called the **maximum value** or

a effective value
b peak value
c basic value
d average value.

33. In Fig. 8 if the sine wave was generated at 50 Hz, the voltage would be extinguished 100 times every

a hour
b minute
c second
d period.

34. The ohm is the unit of all the following terms, *except*

a impedance
b resistance
c reactance
d inductance.

35. If an ac supply of 400 V is allowed to rise by 10% it would become

a 460 V
b 440 V
c 420 V
d 410 V.

36. A **commutator** is part of machine's armature that

a changes alternating current into direct current
b transforms alternating current to different levels
c converts direct current into alternating current
d creates alternating current sinusoidal waveforms.

37. The BS3939 circuit symbol shown in Fig. 9 represents a

- **a** voltage amplifier
- **b** telephone point
- **c** transducer
- **d** pn diode.

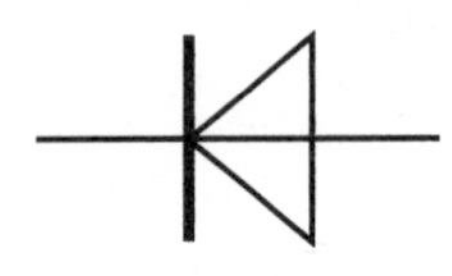

Fig. 9

38. Using the formula $n_s = f/p$, a two-pole ac generator will produce 50 Hz if it is run at a speed of

- **a** 4000 rev/min
- **b** 3000 rev/min
- **c** 2000 rev/min
- **d** 1000 rev/min.

39. Name the electrical equipment that converts ac into dc

- **a** alternator
- **b** transformer
- **c** rectifier
- **d** regulator.

40. Which one of the following materials has a negative temperature coefficient of resistance?

- **a** silver
- **b** aluminium
- **c** carbon
- **d** tin.

41. The ratio

$$\frac{\text{Change of length}}{\text{original length} \times \text{temperature rise}}$$

is called:

- **a** temperature coefficient
- **b** specific heat capacity
- **c** standard atmospheric pressure
- **d** coefficient of linear expansion.

42. Eddy currents are circulating currents found in motors and

- **a** capacitors
- **b** transformers
- **c** resistors
- **d** rectifiers.

43. A double-wound transformer has a primary voltage of 110 V and a secondary voltage of 550 V. If there are 100 turns on the secondary winding the number of turns on the primary winding will be

- **a** 500
- **b** 240
- **c** 50
- **d** 20.

44. In Fig. 10, when the rotating loop is in the position shown, it is induced with

- **a** infinite electromotive force
- **b** maximum electromotive force
- **c** minimum electromotive force
- **d** zero electromotive force.

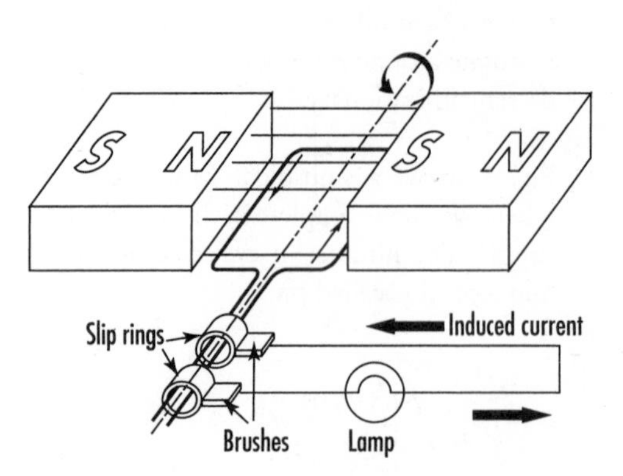

Fig. 10

45. The unit of **magnetic flux** is called the

- **a** joule
- **b** watt
- **c** coulomb
- **d** weber.

46. The unit of electricity is called the

- **a** kilowatt hour
- **b** kilovolt ampere
- **c** ampere hour
- **d** megawatt.

47. The ratio: power output/power input is called

a efficiency
b work done
c turning effort
d horse power.

48. Which one of the following formulae is used for finding electrical energy (W)?

a $W = ECt$
b $W = EIt$
c $W = ER/t$
d $W = EI^2t$.

49. The SI derived unit for **inductance** (L) is the

a joule
b ohm
c henry
d farad.

50. The SI derived unit for **capacitance** (C) is the

a joule
b ohm
c henry
d farad.

ASSESSMENT 2.1

Answers, Hints and References

1. **a** *See* Ref. 6, p 1.
2. **c** *See* Ref. 6, p 15.
3. **c** *See* Ref. 6, p 15.
4. **b** *See* Ref. 6, p 1.
5. **b** *See* Ref. 6, p 9.
6. **d** *See* Ref. 6, p 18, Fig. 2.6.
7. **d** *See* Ref. 6, p 2.
8. **b** *See* Ref. 6, p 20.
9. **c** *See* Ref. 6, p 25.
10. **b** *See* Ref. 6, p 30.
11. **b** *See* Ref. 6, p 22.
12. **c.**
13. **b** *See* Ref. 6, p 31.
14. **a** *See* Ref. 7, p 107.
15. **c** *See* Ref. 6, p 31.
16. **c** *See* Ref. 6, p 33.
17. **c** *See* Ref. 6, p 37, formula 3.11.
18. **d** *See* Ref. 6, p 31.
19. **d** *See* Ref. 6, pp 34–35.
20. **a** *See* Ref. 6, p 3, Table 3.
21. **c** *See* Ref. 6, p 3.
22. **b** *See* Ref. 6, p 36.
23. **a** *See* Ref. 6, p 43.
24. **d** *See* Ref. 7, p 52.
25. **b** *See* Ref. 6, p 39, Fig. 3.14.
26. **c** *See* Ref. 6, p 7, Fig. 1.3.
27. **b** *See* Ref. 6, p 2.
28. **c.**
29. **b** *See* Ref. 7, p 39, Fig. 2.19 and formula 2.16.
30. **b** *See* Ref. 6, p 42, Fig. 3.21.
31. **c** *See* Ref. 6, pp 66–67.
32. **b** *See* Ref. 6, p 67, Fig. 4.29.
33. **c** *See* Ref. 7, p 52.
34. **d** *See* Ref. 6, p 2.
35. **b** $V_{max} = 400 + (400 \times 10/100) = 440$ V.
36. **a** *See* Ref. 6, pp 43–44.
37. **d** *See* Ref. 6, p 61.
38. **b** *See* Ref. 7, p 52.
39. **c.**
40. **c** *See* Ref. 7, p 29.
41. **d** *See* Ref. 6, p 23.
42. **b** *See* Ref. 6, p 45, Fig. 3.28.
43. **d** *See* Ref. 6, p 45.
44. **b** *See* Ref. 7, p 39.
45. **d** *See* Ref. 6, p 2.
46. **a** *See* Ref. 6, p 5.
47. **a** *See* Ref. 6, p 40.
48. **b** *See* Ref. 6, p 36, formula 3.8.
49. **c** *See* Ref. 6, p 2.
50. **d** *See* Ref. 6, p 2.

ASSESSMENT 2.2

1. In Fig. 1, what **force** is needed to lift the load of mass 5 kg, assuming g = 9.81 m/s²?

a 96.24 N
b 49.05 N
c 14.81 N
d 1.962 N.

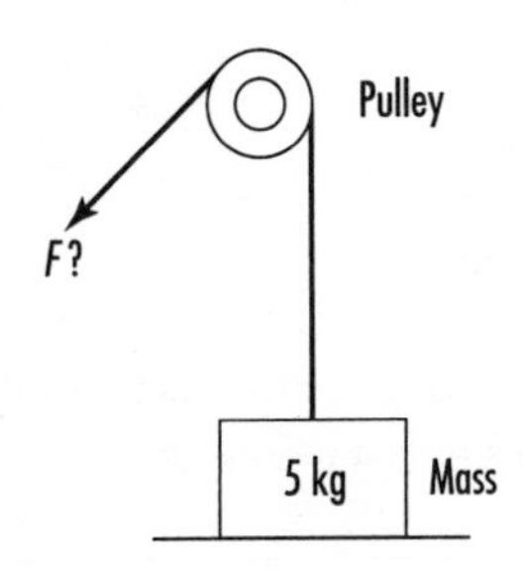

Fig. 1

2. The force needed to stop and start things is called

a inertial force
b friction force
c centrifugal force
d gravitational force.

3. All of the following are basic SI units, *except*

a time
b mass
c length
d area.

4. The unit of **frequency** is called the

a tesla
b weber
c hertz
d farad.

5. In Fig. 2, which object has the more stable equilibrium?

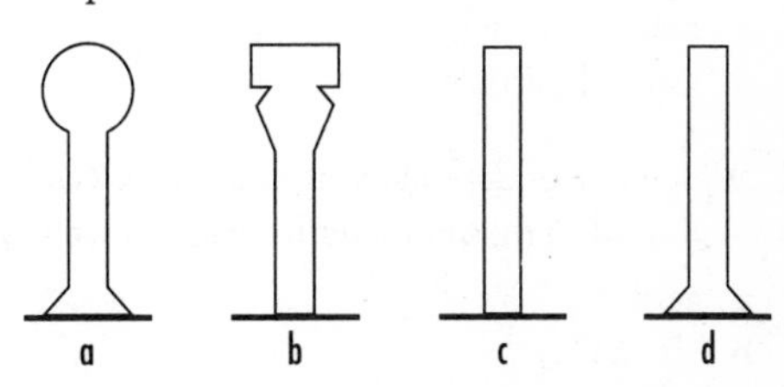

Fig. 2

6. Which one of the following expressions could be used to find **efficiency**?

a output power/input power
b mass/volume
c work done/time taken
d force/distance.

7. Which of the anticlockwise moments (AM) in Fig. 3 will create a balance?

a 3.25 N
b 5.25 N
c 14.5 N
d 21.6 N.

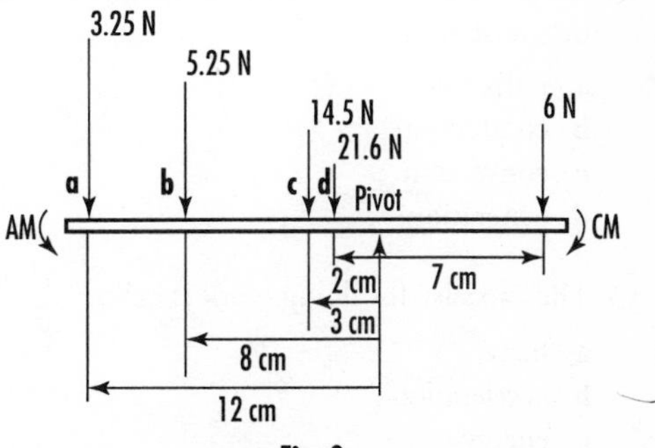

Fig. 3

8. The derived SI unit for **potential difference** is called the

a coulomb
b volt
c kilogram
d ampere.

9. What is 15°C converted into Kelvin?

a 663 K
b 342 K
c 288 K
d 115 K.

10. An electron is a

a negatively charged particle
b positively charged particle
c neutral particle
d ionized particle.

11. Which one of the following substances can be used inside a thermometer instead of mercury

a acid
b alcohol
c milk
d water.

12. What is 30°C converted to degrees Fahrenheit?

a 86°F
b 61°F
c 30°F
d 25°F.

13. A coulomb of electricity is equivalent to

a 11×10^3 volts
b 6.3×10^{18} electrons
c 3.6×10^6 joules
d 1.5×10^{-3} amperes.

14. All the following are ways in which heat can be transferred, *except*

a radiation
b conduction
c convection
d generation.

15. The capacity for doing work is called

a force
b acceleration
c energy
d power.

16. In Fig. 4 the potential difference across R_1 is 20 V while across R_3 it is 50 V. If the supply voltage is 100 V and R_2 is 3 Ω, what is the current taken by the circuit?

a 10 A
b 5 A
c 3 A
d 1 A.

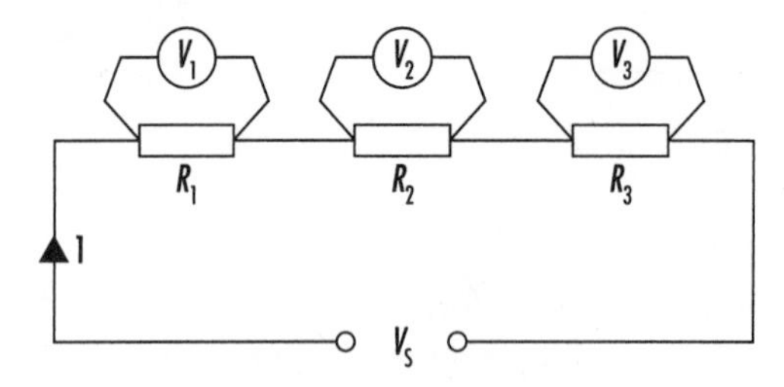

Fig. 4

17. The current taken by a 100 W/230 V tungsten lamp is approximately

a 434.8 mA
b 43.48 mA
c 4.348 mA
d 0.435 mA.

18. All the following are factors that influence a conductor's resistance, *except*

a length
b resistivity
c area
d voltage.

19. In Fig. 5, to obtain a circuit resistance of 46 Ω the following switches need to be closed

a 1, 2, 3 and 4
b 2, 3, 4 and 5
c 3, 4, 5 and 6
d 4, 5, 6 and 7.

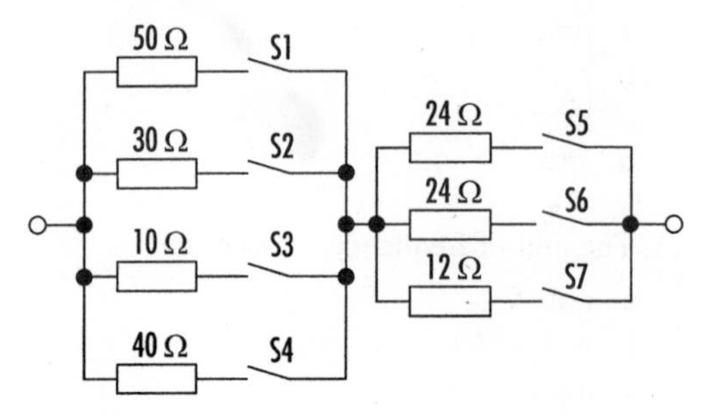

Fig. 5

20. One joule of energy is equivalent to

a 1 ohm metre
b 1 newton metre
c 1 millisecond
d 1 kilowatt hour.

21. Which one of the following metals has the lowest resistivity?

a aluminium
b tungsten
c copper
d brass.

22. A wattmeter has two internal coils, one is a current coil and the other is a

a search coil
b voltage coil
c energy coil
d trip coil.

23. John Fleming devised rules for finding the direction of induced emf magnetic field and motion. The right-hand rule is used for

a generators
b motors
c capacitators
d inductors.

24. The rms value of an ac sine wave is a measure of its usefulness in producing

a a stable maximum value of current
b a current with no ripple effects
c minimum heat loss in conductors
d equal heating effect as direct current.

25. In Fig. 6, what is the power consumed by the wattmeter if A reads 8 A and V reads 250 V?

a 2.00 kW
b 2.00 W
c 2.00 mW
d 2.00 μW.

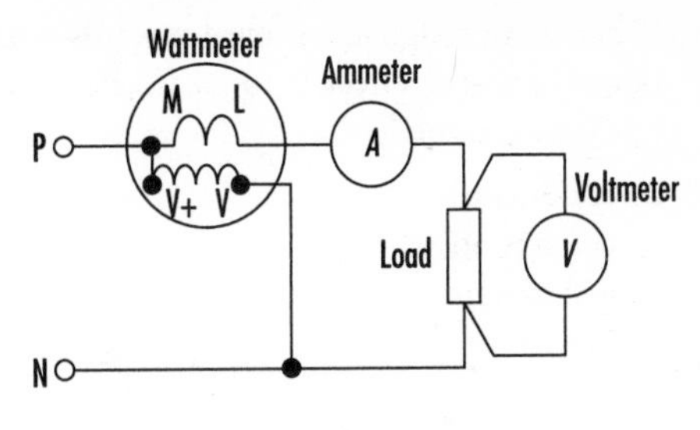

Fig. 6

26. If the charge conveyed in an electric circuit is 300 C, the time taken to pass a current of 10 A is

a 30 ks
b 30 s
c 30 ms
d 30 μs.

27. The SI unit for luminous intensity is the

a lumen
b candela
c lux
d tesla.

28. When two permanent magnets are closely positioned such that their same poles face each other, the effect is for them both to

a attract
b stand on end
c repel
d move away at 90°.

29. In Fig. 7, the type of magnetic field produced by the coil will be

a permanent
b alternating
c stationary
d distorted.

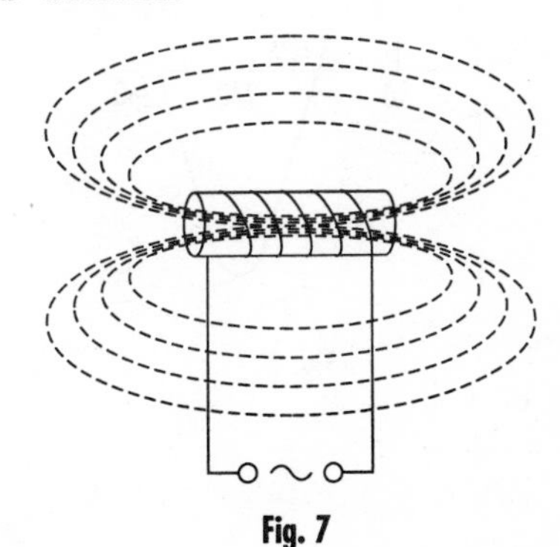

Fig. 7

30. When a bar magnet is moved towards a stationary coil the effect is to create electricity by electromagnetic

a conduction
b conversion
c generation
d induction.

31. What force is exerted on a copper conductor placed at right-angles to a uniform magnetic field of flux density 30 T when its effective length is 500 mm and it carries a current of 5 A?

a 90 N
b 75 N
c 67 N
d 45 N.

32. Fig. 8 shows a sinusoidal voltage quantity taken over one complete cycle (1 Hz) of the supply. The number of times the wave is extinguished over 50 Hz is

a 240
b 150
c 100
d 50.

33. In Fig. 8 the line marked X is determined from the expression

a $V_{max} = V_{rms}/0.637$
b $V_{max} = V_{ave}/0.637$
c $V_{max} = V_{rms}/0.837$
d $V_{max} = V_{ave}/0.707$.

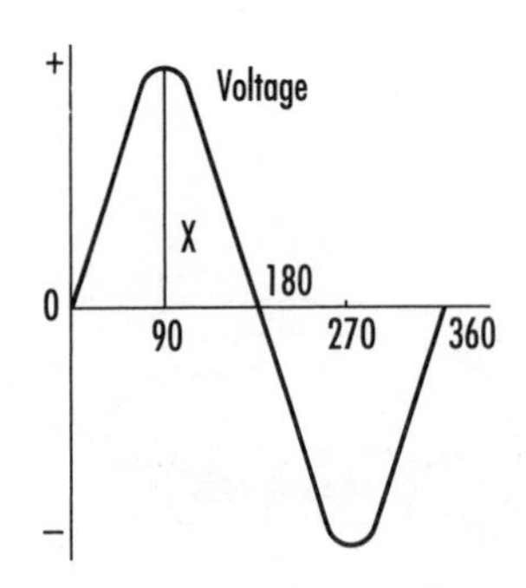

Fig. 8

34. In Fig. 8, if the line marked X had a value of 325.3 V, the rms value of the sine wave would be

a 415 V
b 400 V
c 230 V
d 207 V.

35. In an ac circuit containing resistive and inductive components, the ratio (supply voltage/supply current) is used to find

a energy used
b power consumed
c circuit impedance
d coil reactance.

36. Which one of the following formulae is used for finding **power** (P) consumed in a circuit?

a $P = V^2R$
b $P = I^2R$
c $P = IR^2$
d $P = V^2I$.

37. Which one of the following components is a vital part of a machine's commutator?

a graphite brush
b slip-ring
c shading ring
d cage rotor.

38. The connection of the four diodes shown in Fig. 9 will produce

a full-wave amplification
b full-wave transformation
c full-wave oscillation
d full-wave rectification.

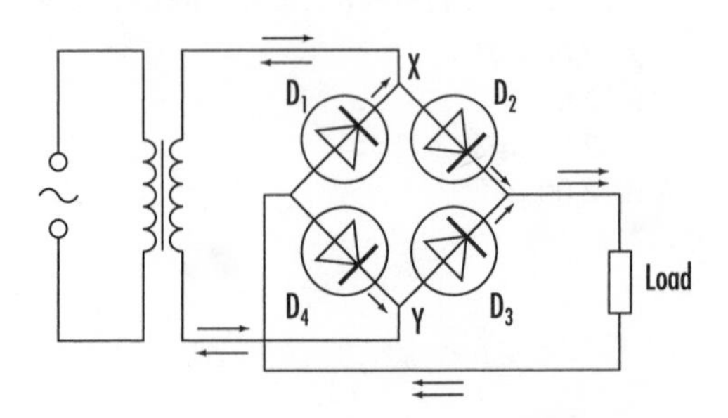

Fig. 9

39. For a two-pole ac generator to produce 50 Hz it must run at a speed of

- **a** 4000 rev/min
- **b** 3000 rev/min
- **c** 2000 rev/min
- **d** 1000 rev/min.

40. Which one of the following is likely to require a smoothing circuit to prevent output current fluctuations?

- **a** variable rheostat
- **b** step-down transformer
- **c** junction diode rectifier
- **d** heating thermostat.

41. Which one of the following materials has a positive temperature coefficient of resistance?

- **a** paraffin
- **b** electrolyte
- **c** annealed copper
- **d** carbon.

42. What is the amount of heat required to raise the temperature of 1.5 litres of water from 20°C to boiling point, given that the specific heat for water is 4200 J/kg°C?

- **a** 630 kJ
- **b** 504 kJ
- **c** 224 kJ
- **d** 126 kJ.

43. Which one of the following components is likely to produce **eddy currents**?

- **a** discharge resistor
- **b** core transformer
- **c** bridge rectifier
- **d** secondary battery.

44. The transformer shown in Fig. 10 is called a three-phase, double-wound

- **a** step-down transformer
- **b** autotransformer
- **c** bar-type transformer
- **d** step-up transformer.

45. In Fig. 10, the windings of the transformer that are normally connected to the high voltage supply are called the

- **a** primary windings
- **b** secondary windings
- **c** tertiary windings
- **d** auxiliary windings.

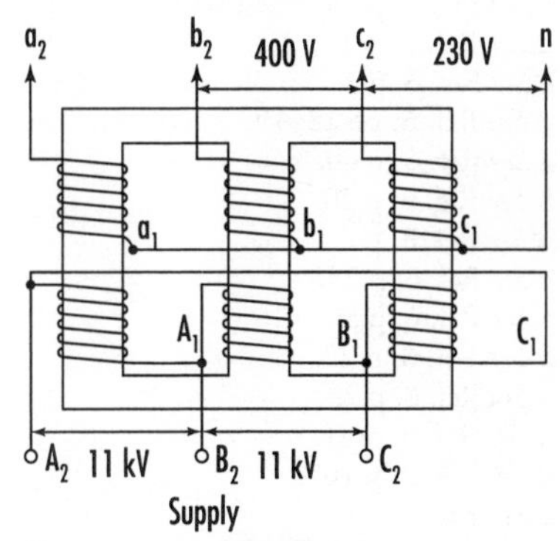

Fig. 10

46. A transformer has a primary to secondary voltage ratio of 100:1. If the secondary voltage supplies 230 V, the primary voltage will be

- **a** 23 MV
- **b** 23 kV
- **c** 23 V
- **d** 23 mV.

47. The SI derived unit for **power** is called the

- **a** joule
- **b** watt
- **c** coulomb
- **d** weber.

48. One kilowatt hour is equivalent to

- **a** 5.0 MJ
- **b** 3.6 MJ
- **c** 2.5 MJ
- **d** 1.0 MJ.

49. The ratio (energy out/energy in) is called

- **a** efficacy
- **b** diversity
- **c** efficiency
- **d** capacity.

50. The farad (F) is the unit of

- **a** reactance
- **b** impedance
- **c** inductance
- **d** capacitance.

ASSESSMENT 2.2

Answers, Hints and References

1. **b** *See* Ref. 6, p 15.
2. **a** *See* Ref. 6, p 15.
3. **d** *See* Ref. 6, p 1.
4. **c** *See* Ref. 6, p 2.
5. **d** *See* Ref. 6, pp 18–19.
6. **a** *See* Ref. 6, p 40.
7. **b** *See* Ref. 6, p 20.
8. **b** *See* Ref. 8, p 2.
9. **c** *See* Ref. 6, p 25.
10. **a** *See* Ref. 6, p 30.
11. **b** *See* Ref. 6, p 22.
12. **a** *See* Ref. 6, p 23.
13. **b** *See* Ref. 6, p 31.
14. **d** *See* Ref. 7, p 107.
15. **c** *See* Ref. 6, p 36.
16. **a** *See* Ref. 6, p 33. The pd across R_2 is 30 V and $I = 30/3 = 10$ A.
17. **a** The question tests your knowledge of prefixes.
18. **d** *See* Ref. 6, p 31.
19. **d** *See* Ref. 6, pp 34–35.
20. **b** *See* Ref. 6, p 26.
21. **c** *See* Ref. 6, pp 31–32.
22. **b** *See* Ref. 7, p 126.
23. **a** *See* Ref. 6, p 43.
24. **d** *See* Ref. 7, p 52.
25. **a** The question tests your knowledge of prefixes.
26. **b** The question tests your knowledge of prefixes. Ref. 6, pp 5, 31.
27. **b** *See* Ref. 6, p 2.
28. **c**.
29. **b** Since the current source is alternating, the magnetic flux will alternate.
30. **d** *See* Ref. 6, pp 42–43.
31. **b** $F = BLI = 30 \times 0.5 \times 5 = 75$ N. *See* Ref. 7, p 39.
32. **c** *See* Ref. 6, p 66.
33. **b** *See* Ref. 6, pp 66–67.
34. **c** *See* Ref. 6, p 67.
35. **c** *See* Ref. 6, p 2.
36. **b** *See* Ref. 6, p 37.
37. **a** *See* Ref. 6, pp 43–44.
38. **d** *See* Ref. 7, p 99, Fig. 5.18.
39. **b** *See* Ref. 7, p 52.
40. **c** *See* Ref. 7, p 99.
41. **c** *See* Ref. 7, p 29.
42. **b** *See* Ref. 6, p 24.
43. **b** *See* Ref. 6, p 45, Fig. 3.28.
44. **a** *See* Ref. 7, p 91, Fig. 5.5.
45. **a** *See* Ref. 7, pp 90–91.
46. **b** *See* Ref. 6, p 45.
47. **b** *See* Ref. 6, p 2, definitions.
48. **b** *See* Ref. 6, p 5.
49. **c** Efficiency $(\eta) = W_0/W_I$
50. **d** *See* Ref. 6, p 2.

ASSESSMENT 2.3

1. What is the unit name for mass?
   - **a** pound
   - **b** stone
   - **c** kilogram
   - **d** tonne.

2. The cause of motion or the effect it produces is called
   - **a** force
   - **b** power
   - **c** energy
   - **d** potential.

3. What is the work done by a crane that lifts a weight of 4000 kg through a height of 12 m? (NB: 1 kg = 9.81 N)
   - **a** 470.88 kJ
   - **b** 48 kJ
   - **c** 4.893 kJ
   - **d** 3.27 kJ.

4. What is the density of air contained in a room of volume 200 m^3 if its mass is 260 kg?
   - **a** 5.2 kg/m^3
   - **b** 2.1 kg/m^3
   - **c** 1.3 kg/m^3
   - **d** 0.8 kg/m^3.

5. The point through which the whole weight of an object seems to act is called the
   - **a** moment of a force
   - **b** centre of gravity
   - **c** centripetal force
   - **d** Newton constant.

6. The expression (force × distance) is called
   - **a** horse power
   - **b** density
   - **c** torque
   - **d** work done.

7. Fig. 1 shows a lever which works on the principle of moments. What is the point marked X on the diagram?
   - **a** zero point
   - **b** fulcrum
   - **c** null point
   - **d** intercept.

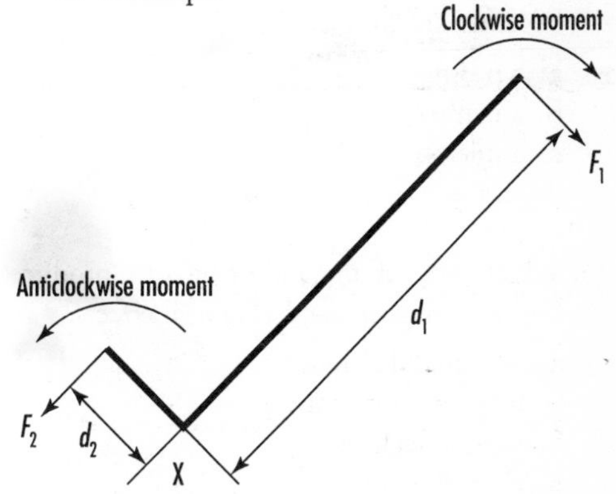

Fig. 1

8. In Fig. 1 what clockwise force is required on the lever to lift a load of 60 kg given that d_1 is 40 cm and d_2 is 4 cm?
   - **a** 30 N
   - **b** 22 N
   - **c** 16 N
   - **d** 6 N.

9. A force of 100 N is required to accelerate a mass of 5 kg at

a 500 m/s^2
b 100 m/s^2
c 20 m/s^2
d 15 m/s^2.

10. Which one of the following is a vector quantity?

a force
b density
c volume
d mass.

11. A thermometer contains mercury because it

a is a poor conductor of heat
b is a good conductor of heat
c has a boiling point of 100°C
d does not expand uniformly.

12. What is 86°F converted to degrees Celsius?

a 50°C
b 42°C
c 30°C
d 25°C.

13. When an atom loses or gains an electron it becomes known as

a a proton
b a neutron
c a micron
d an ion.

14. Which one of the following can produce thermal, magnetic and chemical effects?

a potential difference
b magnetomotive force
c self-inductance
d electric current.

15. The ratio of **potential difference** between the ends of a conductor and the **current** flowing through the conductor is used to find

a resonance
b reactance
c resistance
d resistivity.

16. In Fig. 2 the pd across R_1 is 20 V while across R_3 it is 60 V. If the supply voltage is 100 V and R_2 is 4 Ω, what is the current taken by the circuit?

a 10 A
b 5 A
c 3 A
d 1 A.

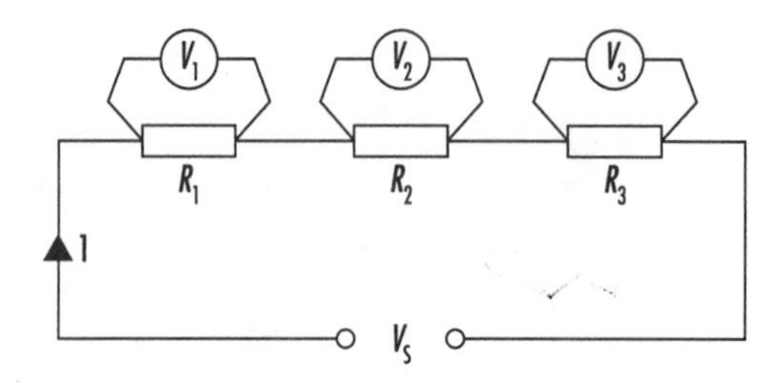

Fig. 2

17. The current taken by a 100 W, 230 V tungsten lamp is approximately

a 0.44 kA
b 43.5 A
c 435 mA
d 4348 μA.

18. What is the resistance of a 10 kW heater taking a current of 25 A?

a 16 Ω
b 10 Ω
c 8 Ω
d 4 Ω.

19. In Fig. 3, to obtain a circuit resistance of 20 Ω the following switches need to be closed

a 1, 2, 3 and 4
b 2, 3, 4 and 5
c 3, 4, 5 and 6
d 4, 5, 6 and 7.

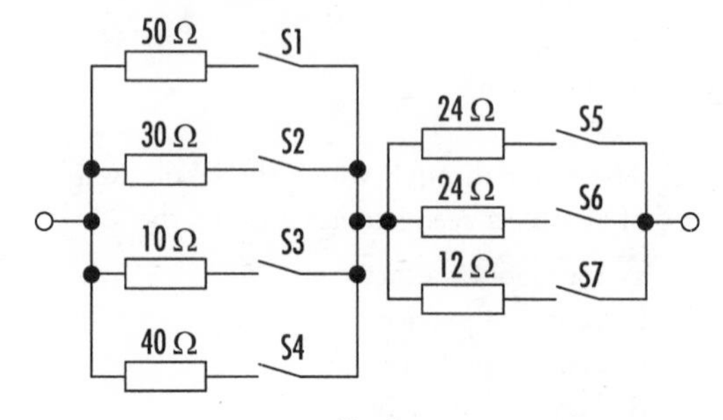

Fig. 3

20. Which one of the following low voltages is *not* normally used in the United Kingdom?

a 500 V
b 400 V
c 230 V
d 110 V.

21. When two similar 230 V/100 W lamps are connected in series with each other across a 400 V supply, the light output from each lamp will be

a brighter
b dimmer
c unchanged
d extinguished.

22. In Fig. 4, which combination of 2 V cells produces the highest output voltage?

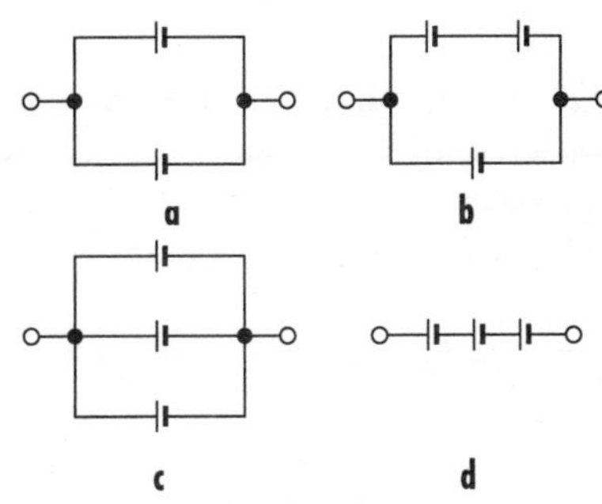

Fig. 4

23. To measure current in a circuit an ammeter is connected

a in parallel with the load
b in series with the load
c across the supply terminals
d in series with a voltmeter.

24. A wattmeter is used in a circuit to measure

a kilowatt hours
b power factor
c consumed power
d potential difference.

25. In Fig. 5, what will W read if A reads 12.5 A and V reads 230 V?

a 661.3 kW
b 35.93 kW
c 18.40 kW
d 2.875 kW.

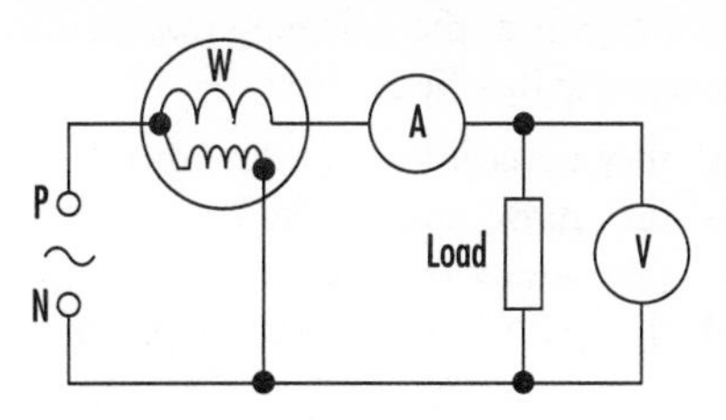

Fig. 5

26. If the charge conveyed in an electric circuit is 300 C, the time taken to pass a current of 3 A is

a 900 s
b 303 s
c 297 s
d 100 s.

27. The derived SI unit for **magnetic flux** is the

a weber
b tesla
c henry
d siemen.

28. Magnetic fields are associated with all the following, *except*

a permanent magnets
b solenoids
c capacitors
d current-carrying conductors.

29. In Fig. 6, which factor will *not* alter the strength of induced emf?

a polarity of conductor
b magnetic field strength
c effective conductor length
d conductor velocity.

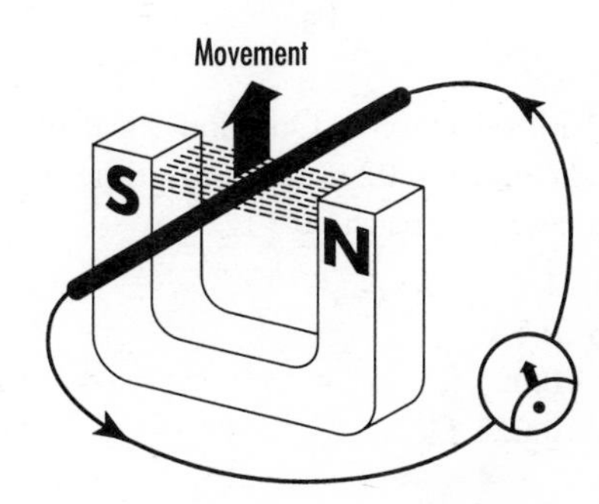

Fig. 6

30. Which one of the following is *not* a rule that **magnetic flux lines** obey?

- **a** they never intersect each other
- **b** they move sinusoidally
- **c** they always try to contract
- **d** they each form a closed loop.

31. What force is exerted on a copper conductor placed at right-angles to a uniform magnetic field of flux density 15 T if its effective length is 250 mm and it carries a current of 20 A?

- **a** 90 N
- **b** 75 N
- **c** 67 N
- **d** 45 N.

32. Fig. 7 shows a sinusoidal voltage quantity taken over one complete cycle (1 Hz) of the supply. For a nominal supply voltage in the United Kingdom, how many cycles are generated?

- **a** 100 Hz
- **b** 75 Hz
- **c** 60 Hz
- **d** 50 Hz

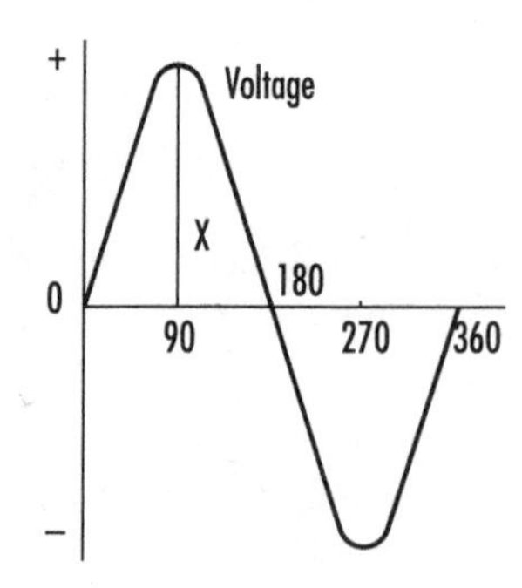

Fig. 7

33. With reference to Question 32, if the periodic time is 0.02 s, what is the time taken for the wave to reach 90°?

- **a** 1 s
- **b** 0.5 s
- **c** 0.005 s
- **d** 0.025 s

34. A standard ac voltage such as 230 V or 400 V is derived from Fig. 7 as

- **a** 0.637 times the basic value
- **b** 0.637 times the effective value
- **c** 0.707 times the maximum value
- **d** 0.707 times the average value.

35. For an ac supply voltage of 230 V, the highest point reached on the sine wave in Fig. 7 is

- **a** 361 V
- **b** 325 V
- **c** 310 V
- **d** 300 V.

36. The term peak-to-peak in a sine wave refers to the distance between

- **a** any two instantaneous points
- **b** the start and finish of each half cycle
- **c** opposite maximum values in one cycle
- **d** the base line and any one maximum value.

37. The **periodic time** (T) taken to complete each cycle of a sine wave is given by the expression

- **a** $T = \pi/f$
- **b** $T = 1/f$
- **c** $T = f/50$
- **d** $T = 2\pi/f$.

38. Which of the following formulae is used for finding current?

- **a** $I = \sqrt{(P/R)}$
- **b** $I = V^2R$
- **c** $I = W/R^2$
- **d** $I = \sqrt{(W/Z)}$

39. The time taken to complete a cycle for an ac supply system operating at 50 Hz is

- **a** 5.00 s
- **b** 1.00 s
- **c** 0.40 s
- **d** 0.02 s.

40. When ac is rectified into dc, it is necessary to smooth the output signal by means of a

- **a** filter circuit
- **b** snubber circuit
- **c** passive circuit
- **d** carrier circuit.

41. Which one of the following is used for rectification purposes?

a double-wound transformer
b semiconductor junction diode
c synchronous motor
d voltage regulator.

42. Which circuit in Fig. 8 will allow both lamps to light?

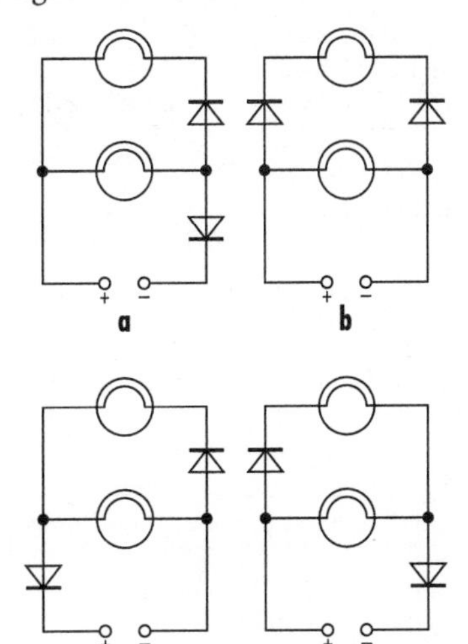

Fig. 8

43. The three vertical members of the transformer core shown in Fig. 9 are called

a yokes
b joiners
c limbs
d struts.

44. The winding connections of the three-phase transformer shown in Fig. 9 are commonly known as

a star–star
b star–delta
c delta–star
d delta–delta.

45. In a double-wound transformer, the winding normally connected to the load is called the

a phase winding
b primary winding
c secondary winding
d tertiary winding.

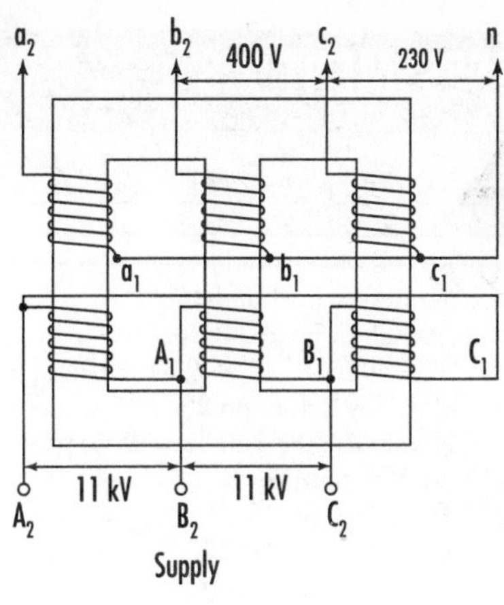

Fig. 9

46. A transformer has a primary to secondary turns ratio of 25:1. If the primary voltage is 250 V then the secondary voltage will be

a 40 V
b 30 V
c 20 V
d 10 V.

47. The unit of **energy** is called the

a joule
b watt
c coulomb
d weber.

48. Ten kilowatt hours is equivalent to

a 50 MJ
b 36 MJ
c 25 MJ
d 10 MJ.

49. What is the efficiency of a 3 kW/240 V electric heater if it takes a current of 10.5 A at 230 V?

a 92.5%
b 80.5%
c 56.4%
d 34.9%.

50. In SI derived units, the henry (H) is the unit of

a impedance
b capacitance
c inductance
d reactance.

ASSESSMENT 2.3

Answers, Hints and References

1. **c** *See* Ref. 6, pp 1 and 15.
2. **a** *See* Ref. 6, pp 2 and 16.
3. **a** $W = 4000 \times 12 \times 9.81 \times 0.001 = 470.88$ kJ.A. See Ref. 6, pp 25–26.
4. **c** ($p = 260/200 = 1.3$) *See* Ref. 6, p 21.
5. **b** *See* Ref. 6, p 19.
6. **d** *See* Ref. 6, pp 25–27.
7. **b** *See* Ref. 6, pp 21–22.
8. **d** $F_1 = (60 \times 4)/40 = 6$ N.
9. **c** ($a = 100/5$) *See* Ref. 6, p 17, formula 2.5.
10. **a** *See* Ref. 6, pp 16–18.
11. **b** *See* Ref. 6, p 22.
12. **c** (°C =) $5/9 \times (86 - 32) = 30$°C
13. **d** *See* Ref. 6, p 30.
14. **d** *See* Ref. 6, pp 36–50.
15. **c** *See* Ref. 6, p 31.
16. **b** The current will be the same through each resistor. The pd across R_2 is 20 V, therefore $I = 20/4 = 5$ A.
17. **c** The question tests your knowledge of prefixes.
18. **a** $R = P/I^2 = 10\,000/6125 = 16\ \Omega$.
19. **c** *See* Ref. 6, p 35.
20. **a** Remember that **b** and **c** are nominal voltages.
21. **b** The voltage for each lamp is now only 200 V.
22. **d** The cell voltages are added together.
23. **b** They must have a low resistance to allow all the circuit current to be measured.
24. **c** *See* Ref. 6, p 37.
25. **d** $P = VI = 230 \times 12.5 = 2875$ W.
26. **d** *See* Ref. 6, pp 5 and 31.
27. **a** *See* Ref. 6, p 2.
28. **c** *See* Ref. 6, p 56 or Ref. 7, p 41.
29. **a** *See* Ref. 6, p 42.
30. **b** *See* Ref. 6, p 42.
31. **b** ($F = BLI = 15 \times 0.25 \times 20 = 75$ N).
32. **d** *See* Ref. 7, p 52.
33. **c** *See* Ref. 6, pp 66–67.
34. **c** *See* Ref. 6, p 67.
35. **b** $V_{max} = 230/0.707 = 325$ V.
36. **c** *See* Ref. 6, p 67, Fig. 4.29.
37. **b** *See* Ref. 6, p 66.
38. **a** *See* Ref. 6, p 5.
39. **d** ($T = 1/f = 1/50 = 0.02$ s), see Ref. 1, p 66.
40. **a** *See* Ref. 7, pp 98–99.
41. **b** *See* Ref. 6, pp 59–61.
42. **d** Assume conventional current flow and the diodes pointing in the direction shown.
43. **c** *See* Ref. 6, p 46, Fig. 3.29.
44. **c** *See* Ref. 7, p 9 1, Fig. 5.5.
45. **c** *See* Ref. 7, pp 90–91.
46. **d** *See* Ref. 6, p 45.
47. **a** *See* Ref. 6, p 2, definitions.
48. **b** *See* Ref. 6, p 5.
49. **b** Efficiency = $P_O/P_I = 2.875/3 = (80.5\%)$.
50. **c** *See* Ref. 6, p 2, definitions.

3

OBSERVING SAFETY PRACTICES AND PROCEDURES

To tackle the assessments in Topic 3 you will need to know:

- the purpose and scope of listed statutory and non-statutory regulations governing the control of safety at work;
- employer and employee responsibilities towards the maintenance of safety and provision of a safe working environment;
- employee responsibilities towards the customer and general behaviour expected on site with respect to authorized visitors;
- the correct procedure for the reporting of accidents as well as the environmental conditions leading to accidents in the workplace;
- the procedure for the treatment of accidents, particularly electric shock and burns, and the precautions needed to reduce the risks;
- the procedure for a safe system of work with respect to isolation of 'live' and 'dead' circuits;
- the requirements for notices and labelling of circuits and the types of safety signs and symbols used to promote safety at work;
- the general rules for the observance of safety practices, and identification of suitable access equipment for specified situations.

DEFINITIONS

Appointed person – a person authorized by an employer to take charge of first-aid arrangements (*see* Ref. 8).
Arm's reach – a zone of accessibility to touch with the bare hands without any assistance (*see* Ref. 4, definitions).
Competent person – a person who has the experience and technical knowledge to carry out a specific task(s) or is working under supervision (*see* Ref. 9).
Dead – at or about earth potential and disconnected from any live system.
Direct contact – contact of persons or livestock with live parts which may result in an electric shock (*see* Ref. 4, definitions).
Earth clamp – an electrical accessory used to maintain earth continuity between two conductive parts.

Earth clamp

Extra-low voltage – a voltage which normally does not exceed 50 V ac or 120 V dc whether between conductors or to earth (*see* Ref. 4, definitions).
First aider – a person who is trained in first-aid and who holds a first-aid certificate.
Index of protection – a code system that uses numerals to indicate the degree of protection of persons against contact with live parts inside enclosures and also protection of equipment against ingress of water.
Indirect contact – contact of persons or livestock with exposed conductive parts which have become live under fault conditions.
Isolation – a means to ensure that the supply remains switched off and cannot be inadvertently reconnected.
Mouth-to-mouth ventilation – a method of resuscitating a victim if he/she has stopped breathing.

Permit title	Permit number
Job location	
Plant identification	
Description of work to be done	
Hazard identification	
Precautions necessary	
Protective equipment	
Authorisation	
Acceptance	
Extension/hand over procedures	
Hand back procedures	
Cancellation	

Permit-to-work form

Permit-to-work form – a safety procedure which specifies the work to be done and the precautions to be taken.
Prohibition notice – a document served on an employer who has contravened a legal requirement under the ***Health and Safety at Work Act*** in order to immediately stop a work activity.
Regulations – these are either statutory (legally binding) or non-statutory (codes of practice). Some examples of Statutory Regulations are:

(i) ***Health and Safety at Work Act 1974*** provides the legislative framework to promote, stimulate and encourage high standards of health and safety at work.
(ii) ***Electricity at Work Regulations 1989*** requires precautions to be taken against the risk of death or personal injury from electricity in work activities.
(iii) ***Reporting of Injuries, Diseases and Dangerous Occurrences Regulations (RIDDOR) 1985*** requires injuries, diseases and occurrences in specified categories to be notified to the relevant enforcing authority.

Some examples of non-statutory regulations and other safety publications are:

(i) ***IEE Wiring Regulations 1992*** (BS7671) designed to protect persons, property and livestock against the hazards arising from electrical installations.
(ii) ***HSE Guidance Note GS 27 Protection against electric shock.***
(iii) ***IEE Guidance Note 2 Isolation & Switching*** (2nd edition).

Safe system of work – a formal procedure which results from systematic examination of a task in order to ensure that hazards are eliminated or risks minimized.
Safety policy statement – a written, legal statement by an employer who employs five or more people which sets out the general policy of protecting their health and safety at work.

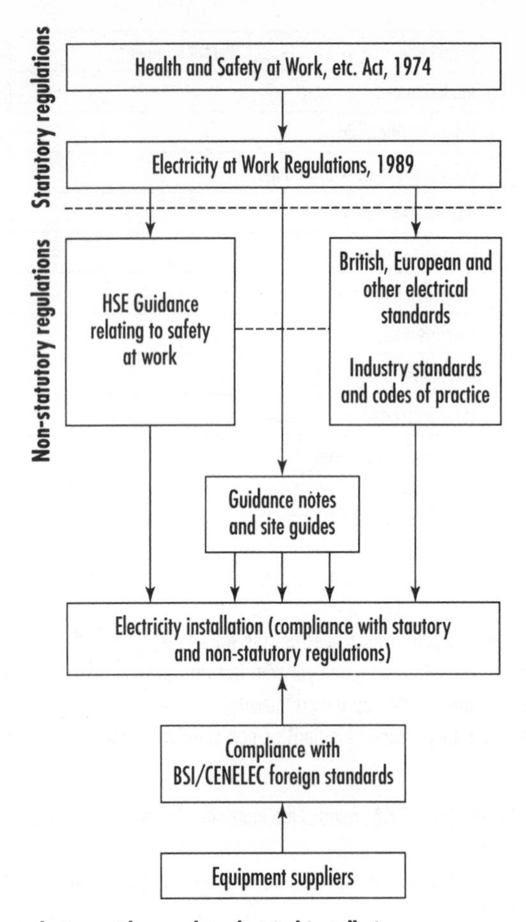

Statutory and non-statutory regulations with regard to electrical installation

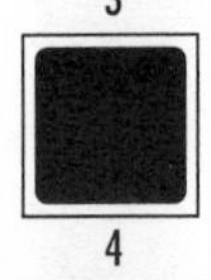

Safety signs

Category
1. Prohibition
2. Warning
3. Mandatory first aid/ safe condition
4. Emergency/escape/ first aid/ safe condition
5. Fire equipment

Meaning
1. Do not do
2. Danger
3. Must do
4. The safe way
5. Location or use of fire equipment

DANGER
415 volts

Note
Prohibition sign – is a red circle on a white background
Warning sign – is a yellow triangle with a black border on a white background
Mandatory sign – is a blue square/rectangle on a white background
Emergency sign – is a green square/rectangle on a white background
Fire Equipment – is a red square/rectangle on a white background

TOPIC 3

Observing Safety Practices and Procedures

ASSESSMENTS 3.1 – 3.3

Time allowed: 1½ hours

Instructions

* You should have the following:

  Question Paper
  Answer Sheet
  HB pencil
  Metric ruler

* Enter your name and date at the top of the Answer Sheet.
* When you have decided a correct response to a question, on the Answer Sheet, draw a straight line across the appropriate letter using your HB pencil and ruler (see example below).
* If you make a mistake with your answer, change the original line into a cross and then repeat the previous instruction. There is only one answer to each question.
* Do not write on any page of the Question Paper.
* Make sure you read each question carefully and try to answer all the questions in the allotted time.

Example:

a 400 V	**a** 400 V
~~**b**~~ 315 V	~~**b**~~ (crossed) 315 V
c 230 V	**c** 230 V
d 110 V	~~**d**~~ 110 V

ASSESSMENT 3.1

1. Which one of the following is a non-statutory regulation?

- **a** Electricity at Work Regulations 1989
- **b** IEE Wiring Regulations 1992
- **c** Electricity Supply Regulations 1988
- **d** Electrical Equipment (Safety) Regulations 1994.

2. Which one of the following Acts of Parliament addresses all people at work and imposes duties on employers and employees?

- **a** Environmental Protection Act 1990
- **b** Mines and Quarries Act 1954
- **c** Health and Safety at Work Act 1974
- **d** Consumer Protection Act 1987.

3. The *Electricity at Work Regulations* are concerned with

- **a** securing the health, safety and welfare of people at work
- **b** safety and comfort of all occupants who use electricity at work
- **c** risk of death and personal injury from electricity used in work
- **d** implementing electricity supplies in all work situations.

4. Which one the following is *not* a function of a works safety representative?

- **a** investigating dangerous occurrences
- **b** investigating complaints by employees
- **c** carrying out work inspection of premises
- **d** writing reports for HSE personnel.

5. Name the meaning of the mandatory safety sign shown in Fig. 1

- **a** wear overalls
- **b** pedestrian route
- **c** personal protection
- **d** wear protective clothing.

Fig. 1

6. The statutory instrument *RIDDOR* is concerned with

- **a** the reporting of injuries and diseases at work
- **b** electrical wiring and equipment used at work
- **c** welfare services of young people at work
- **d** health, safety and welfare in workshops.

7. One type of notice that a Health and Safety Executive (HSE) inspector can serve on an employer who contravenes a requirement of the *Health and Safety at Work (HSW) Act* is

- **a** a disclosure notice
- **b** an improvement notice
- **c** an incident notice
- **d** a danger notice.

8. Which one of the following tools is likely to develop a mushroom head if constantly used?

a cold chisel
b screwdriver
c pein hammer
d bradawl.

9. Which one of the following safety signs in Fig. 2 could be applicable in a work situation involving flying fragments?

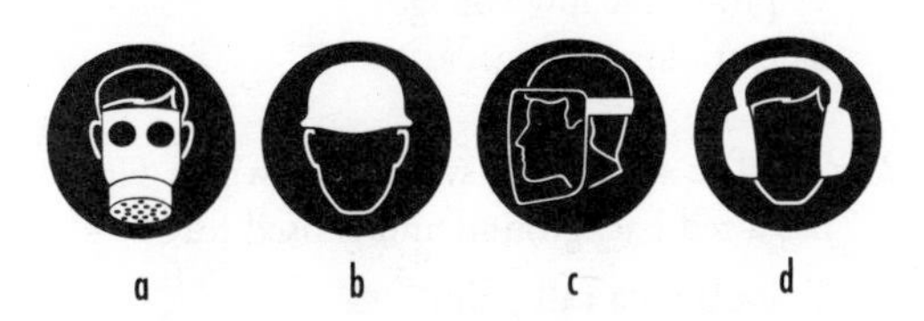

Fig. 2

10. On a building site, a safety helmet must be worn by everyone, *except* where

a attendance on site is less than 1 hour
b a project meeting is held in a site hut
c the work does not exceed 10 m in height
d the operatives are working above ground.

11. Emergency first-aid treatment is a procedure which does not include

a preservation of life
b prevention of the condition worsening
c causation of the accident
d promotion to recover.

12. All the following are recognized solvents but *not*

a vaseline
b acetone
c white spirit
d petroleum spirit.

13. The symbol shown in Fig. 3 is used for labelling substances that are

a irritant
b corrosive
c harmful
d toxic.

Fig. 3

14. With reference to Fig. 4, the object of placing an unconscious person in the recovery position is to

a reduce the possibility of shock
b encourage the heart to beat
c help maintain an open airway
d make the victim nauseous.

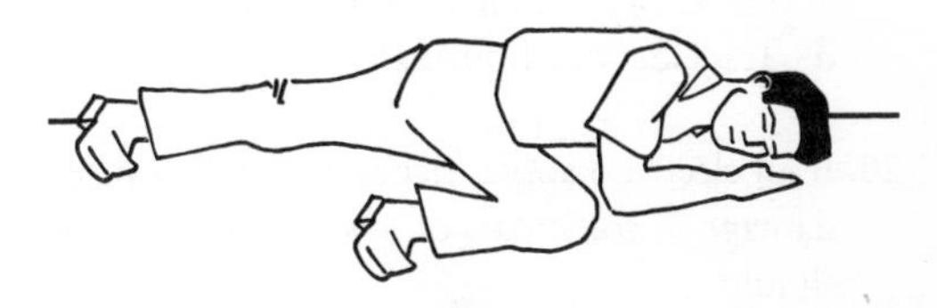

Fig. 4

15. An employer is fulfilling the requirements of the law if he issues his employees with a relevant

a NICEIC newsletter
b HSE approved code of practice
c BSI code of practice
d IEE guidance notes.

16. Which item below is *not* essential in a first-aid box?

a sterile dressing
b sponge
c triangular bandage
d eye pad.

17. All the following are important points to consider if attending a person in shock after a sudden accident, *except*

a loosen clothing around neck
b keep reasonably warm
c treat serious injuries
d give sweet hot tea to drink.

18. If a fellow workmate suffers burns on his hand as a result of using a soldering iron, one of the important things to do is

a remove any blisters with a wet salt cotton wool swab
b remove clothing that has stuck to the wound
c quickly place the wounded part in cold running water
d apply soothing ointment or lotion to the wounded part.

19. Which one of the following organizations publishes specific guidance notes on the requirements of BS7671 (16th edition)?

a Institution of Electrical Engineers
b City & Guilds of London Institute
c Electricity Training Association
d Trade Service Information.

20. If an electrical apprentice accidentally causes damage to some part of a wiring system, he should

a try and repair it himself
b report it to his employer
c report it to his supervisor
d ignore it if it is not too bad.

21. When using a knife to remove the outer sheath from a cable, you should make sure that you cut the insulation

a only after it has been ringed and scored
b only after it has been softened with a heat source
c with the blade pointing towards your body
d with the blade pointing away from your body.

22. The immediate procedure for dealing with a cut finger that is bleeding profusely is to

a seek assistance from a first-aider
b place the wound under a cold tap
c send or phone for an ambulance
d apply direct pressure to the wound.

23. Contact of persons or livestock with 'live' parts which may result in electric shock is defined in the *IEE Wiring Regulations* as

a accidental contact
b indirect contact
c direct contact
d open contact.

24. An ac voltage that does *not* normally exceed 50 V is known as

a safety low voltage
b functional low voltage
c protective low voltage
d extra-low voltage.

25. The fire safety sign symbol shown in Fig. 5 has a red background and is used for

a locating a call point
b starting a water sprinkler
c opening an inlet valve
d stopping a sounder.

Fig. 5

26. Which one of the following pipeline colours is used for electrical services and ventilation ducts?

a orange
b green
c black
d silver.

27. 'The means to ensure that the electricity supply is switched off and cannot be inadvertently switched back on again', is defined in the *IEE Wiring Regulations* as

a isolation
b disconnection
c separation
d disengagement.

28. A safety precaution when using a pillar drill is to

a keep within reach of the emergency stop button
b not wear a wrist watch, jewellery or rings
c wear safety footwear and a safety helmet
d keep long hair under a hair-net or a hat.

29. With reference to the BS1363 13 A plug in Fig. 6, name the type of British Standard fuse that should be fitted

a BS88
b BS1361
c BS1362
d BS3036.

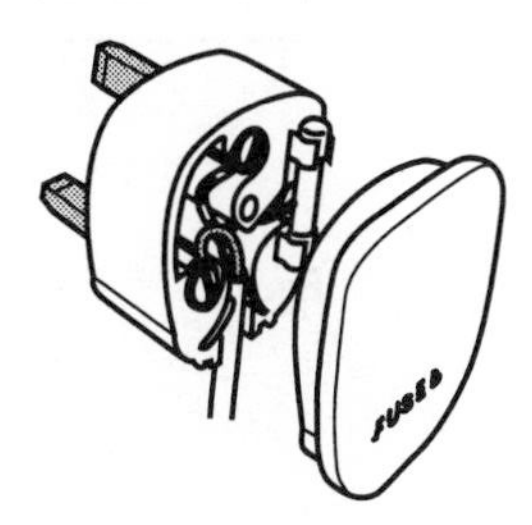

Fig. 6

30. Which one of the following is incorrect when terminating the flex cores inside the accessory shown in Fig. 6?

a blue insulated conductor wired to the neutral terminal
b green/yellow insulated conductor to be cut out if a Class I appliance
c conductor insulation to be left bare 5 mm from each terminal post
d conductors to be placed on the terminal screws in a clockwise direction.

31. What safety form is to be used to protect employees working in hazardous areas?

a work acceptance form
b permit to work form
c dangerous work form
d accident at work form.

32. When working on a live electrical system, one of the precautions to take is having

a some control of the working area
b use of suitable test equipment
c verbal confirmation that the supply is off
d staff with mixed ability and competence.

33. The term 'proving **dead**' at the point of working means checking

a to see if protective devices have been switched off or removed
b with other people that the circuit or system is completely safe
c to see if test instruments are capable of reading accurately
d with a reliable tester to see if the electricity supply remains off.

34. Which one of the following is most likely to give a person a harmful electric shock if accidentally touched?

a neutral of an unbalanced system
b live conductor of a SELV system
c bare earth electrode of a TT system
d live enclosure of an earthed system.

35. Which one of the following safety signs in Fig. 7 should be displayed in a building continually producing obnoxious vapours?

Fig. 7

36. In Fig. 8, the words that are inscribed on the BS951 earth clamp should read:

- **a** Earth Terminal Connection – Do Not Remove
- **b** Safety Electrical Connection – Do Not Remove
- **c** Electrical Earth Connection – Do Not Disconnect
- **d** Earth/Bonding Connection – Do Not Disconnect.

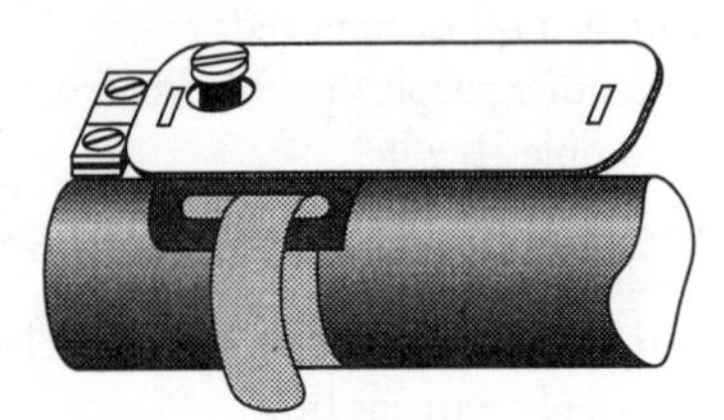

Fig. 8

37. When work is carried out on a final circuit that has been made '**dead**' the warning sign in Fig. 9 should be attached to the

- **a** main intake position
- **b** place of working
- **c** local fuseboard
- **d** last isolation point.

Fig. 9

38. Which one of the following need *not* be present to start a fire?

- **a** oxygen
- **b** carbon
- **c** fuel
- **d** heat.

39. For the residual current device shown in Fig. 10 to protect a person from electric shock, its trip coil should operate at

- **a** 500 mA
- **b** 300 mA
- **c** 100 mA
- **d** 30 mA.

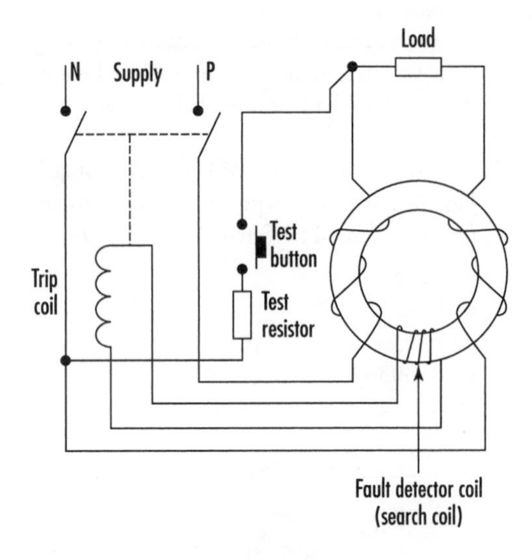

Fig. 10

40. The instructions printed on a residual current device require it to be tested

- **a** every 2 years
- **b** every year
- **c** quarterly
- **d** weekly.

41. Which one of the following safety protective items is required to be worn when chipping hard stone with a hammer and chisel?

- **a** goggles
- **b** helmet
- **c** gloves
- **d** shoes.

42. The *IEE 'On-Site Guide'* recommends a diagram, chart or schedule to show the

- **a** method of providing protection against direct contact
- **b** operation of residual current devices and miniature circuit breakers
- **c** calculation of prospective short circuit current
- **d** number of points, sizes/types of cable for circuits.

43. Which one of the following is *not* a preventative measure to reduce electric shock?

- **a** enclosure of live conductors
- **b** earthing and bonding of pipes
- **c** use of double insulated tools
- **d** circuits with fuses and miniature circuit breakers.

44. Which one of the following fire extinguishers should not be used on burning liquids?

- **a** water
- **b** halon
- **c** foam
- **d** carbon dioxide.

45. Some general rules for observing safe practices on site involve being alert and perceptive, knowing emergency procedures and also

- **a** knowing trade union rules
- **b** being aware of visitors
- **c** understanding contract law
- **d** protecting oneself and others.

46. Which one of the following is likely to be the reason for evacuating people from a building?

- **a** thunder and lightning
- **b** terrorist activity
- **c** management meeting
- **d** electricity failure.

47. When resting a long ladder against a landing place (*see* Fig. 11), it should be set at the correct angle and be

- **a** extended by at least 1.5 m
- **b** properly secured by lashing
- **c** free from ground vibration
- **d** labelled with a danger notice.

48. All of the following are safety points whilst working on a ladder except

- **a** wearing of a safety helmet
- **b** body kept between stiles
- **c** wearing of proper footwear
- **d** unnecessary over-reaching.

Fig. 11

49. The part of a ladder that protects the user if a rung breaks is called a

- **a** reinforcer
- **b** cross bar
- **c** tie rod
- **d** safety wire.

50. The pupose of the guard rail in Fig. 12 is to

- **a** prevent the user from falling
- **b** prevent the tower touching live parts
- **c** secure the top of the ladder
- **d** allow other sections of scaffolding to be added.

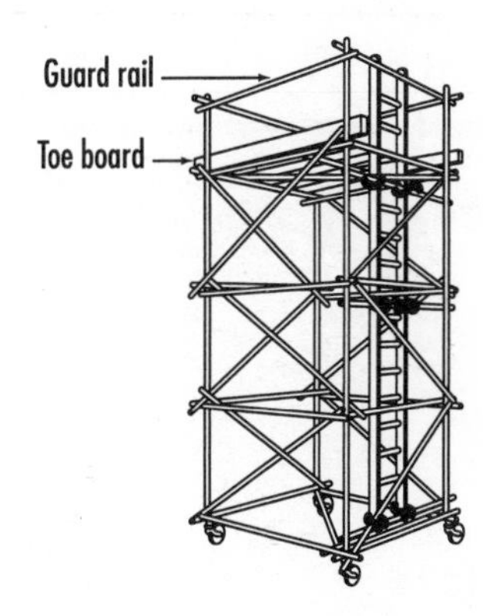

Fig. 12

ASSESSMENT 3.1

Answers, Hints and References

1. **b** *See* Ref. 1, p 18.
2. **c** *See* Ref. 1, pp 12–13.
3. **c** *See* Ref. 1, p 13.
4. **d** *See* Ref. 1, pp 13–15.
5. **b** *See* Ref. 11.
6. **a** *See* Ref. 1, p 15.
7. **b** *See* Ref. 1, p 12.
8. **a** *See* Ref. 12, p 23.
9. **c** *See* Ref. 11.
10. **b** *See* Ref. 1, p 18.
11. **c** *See* Refs 13 and 14.
12. **a** *See* Refs 15 and 16.
13. **d** *See* Ref. 11.
14. **c** *See* Ref. 1, pp 23–24.
15. **b** *See* Ref. 17, section 7.
16. **b** *See* Ref. 1, pp 20–21 and Ref. 13.
17. **d** *See* Ref. 1, p 23.
18. **c** *See* Ref. 1, p 21.
19. **a** *See* Ref. 18.
20. **c** This is the person supervising training.
21. **d** *See* Ref. 12, p 23.
22. **d** *See* Ref. 12, p 26.
23. **c** *See* Ref. 4, definitions.
24. **d** *See* Ref. 1, p 22.
25. **a** *See* Ref. 11.
26. **a** *See* Ref. 11.
27. **a** *See* Ref. 4, definitions.
28. **d**.
29. **c** *See* Ref. 19, pp 13–14.
30. **c** *See* Ref. 19, pp 13–14.
31. **b** *See* Ref. 1, p 131, answer 3.
32. **b** *See* Ref. 1, p 115 and Ref. 10, regulation 14.
33. **d** *See* Ref. 20, pp 9–11.
34. **a** It is a live conductor carrying current.
35. **b** *See* Ref. 11.
36. **b** *See* Ref. 19, p 92, Fig. A1.8.
37. **c** *See* Ref. 20, p11.
38. **b** *See* Ref. 1, pp 32–33.
39. **d** *See* Ref. 1, pp 22–23.
40. **c** *See* Ref. 4, regulation 514–12–02.
41. **a**.
42. **d** *See* Ref. 4, regulation 514–09–01.
43. **d** Their sole function is to protect circuits.
44. **a** *See* Ref. 1, p 33.
45. **d**.
46. **b**.
47. **b** *See* Ref. 1, p 30, Fig. 2.15.
48. **a**.
49. **c** *See* Ref. 21, p 62, book 4.

50 **a** *See* Ref. 1, p 30.

ASSESSMENT 3.2

1. The *Electricity at Work Regulations* came into force in 1990 and replaced the
   - **a** Electricity (Factory Act) Regulations
   - **b** Electricity Supply Regulations
   - **c** Electrical Equipment (Safety) Regulations
   - **d** Electricity Act.

2. Information, instruction, training and supervision are employer duties under the
   - **a** Explosives Act
   - **b** Health and Safety at Work Act
   - **c** Electricity Act
   - **d** Factories Act.

3. The type of 'switching' defined in the *IEE Wiring Regulations* to give normal operation of electrical equipment is called
   - **a** sub-circuit switching
   - **b** maintenance switching
   - **c** emergency switching
   - **d** functional switching.

4. An employer who issues an employee with a relevant HSE Approved Code of Practice to support his instructions is
   - **a** departing from his professional duties
   - **b** removing himself from any accident liability
   - **c** fulfilling the requirements of the law
   - **d** making work activities easier to complete.

5. The *HSW Act* applies to all the following, but *not*
   - **a** public servants in domestic premises
   - **b** chief executives of large companies
   - **c** self-employed people under contract
   - **d** civil servants and medical personnel.

6. A formal safe procedure that involves assessing, implementing and monitoring a task or some maintenance activity is called a
   - **a** variation order
   - **b** permit-to-work
   - **c** work voucher
   - **d** commencement note.

7. Which one of the following, made under the *HSW Act*, is an employee's duty to his/her employer?
   - **a** co-operate in complying with a non-statutory requirement
   - **b** work excess weekly hours outside of the normal contract
   - **c** make arrangements for his or others' safe journey to work
   - **d** take reasonable care to avoid injury to himself and others.

8. Name the term that is defined in the *IEE Wiring Regulations* as a 'zone of accessibility to touch with the bare hands without any assistance'
   - **a** equipotential distance
   - **b** simultaneous part
   - **c** restrictive location
   - **d** arm's reach.

9. In Fig. 1, which electrical item is *not* a method of isolation?

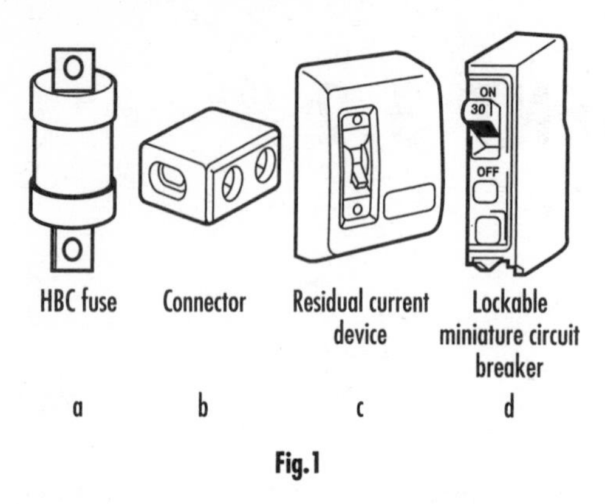

Fig. 1

10. The warning symbol shown in Fig. 2 is used for labelling substances that are

a an irritant
b corrosive
c highly toxic
d non-poisonous.

Fig. 2

11. The *Noise at Work Regulations* requires employers to provide ear protection notices when noise levels are at

a 90 dB(A)
b 70 dB(A)
c 60 dB(A)
d 45 dB(A).

12. The safety helmet sign in Fig. 3 falls in the category of a

a warning sign
b mandatory sign
c prohibition sign
d safe condition sign.

Fig. 3

13. The method of classifying and labelling electrical equipment against the entry of liquids and solids is known as

a material classification
b environment listing
c index of protection
d material indexing.

14. Under the *Health and Safety at Work Act*, where an employer employs five or more employees, he has to provide a

a safety policy statement
b register of accidents
c permanent safety officer
d code of good conduct.

15. Which safety symbol in Fig. 4 is used for giving a general warning against risk of danger?

Fig. 4

16. When a visitor attends a meeting on a large building site for the first time, he/she should preferably be made aware of the site's

a safety rules and accident procedures
b previous history and present planning
c subcontractors' working accommodation
d facilities for car parking and storage.

17. The type of electrolyte used inside a modern Plante lead–acid battery is called

a copper sulphate
b pure sulphuric acid
c sodium chloride
d hydrochloric acid.

18. Which item listed below is not a requirement for a first-aid box?

a treatment notes
b Dettol
c triangular bandage
d sterile dressing.

19. Which one of the following environmental conditions is most likely to lead to an accident in the workplace where revolving machinery is used?

a lighting that produces stroboscopic effects
b ventilation from high-bay industrial fan heaters
c noise that persists between 50 and 70 decibels
d constant activity from overhead gantry cranes.

20. In Fig. 5, the **Index of Protection** code shown on the residual current device/socket outlet signifies that it is protected against

a corrosion and condensation
b chemicals and hot temperatures
c dust and powerful jets of water
d large objects and dripping water.

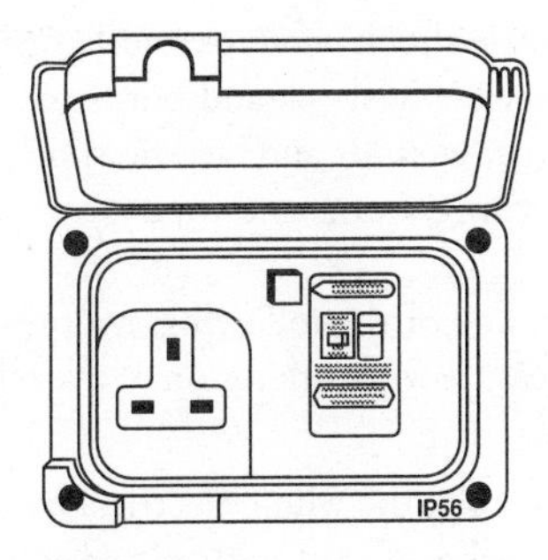

Fig. 5

21. Which one of the following articles of protective clothing is less likely to be considered for wearing while installing heavy armoured cables?

a industrial shoes
b overalls/boiler suit
c industrial gloves
d safety helmet.

22. Which one of the following lamps, when broken in a dry environment, can release toxic mercury and cadmium dust?

a tungsten filament lamp
b low pressure sodium lamp
c tungsten halogen lamp
d low pressure fluorescent lamp.

23. If you discover a fault on a piece of electrical equipment that you are using at work, which person listed below would you notify about the problem?

a safety officer
b product manufacturer
c immediate supervisor
d safety representative.

24. The residual current device shown in Fig. 5. is designed to protect the 'user' against the danger of

a a short circuit
b an open circuit
c a current surge
d an earth fault.

25. Which one of the following lamps, when broken in a recognized lamp crusher, requires soaking in water to reduce the possibility of fire?

a tungsten filament lamp
b low pressure sodium lamp
c tungsten halogen lamp
d low pressure fluorescent lamp.

26. Which hand tool listed below should be used in the direction towards the body of the user to avoid injury?

a spanner
b file
c screwdriver
d pein hammer.

27. The meaning of the safety sign shown in Fig. 6 is

a do not touch
b in this direction
c wear gloves
d beware of hand traps.

Fig. 6

28. Which one of the following can be excluded as a basic precaution during the actual use of a portable electric drill?

a checking that the working voltage is correct
b keeping the lead away from the work area
c being aware when the bit pierces the hole
d avoiding additional pressure on the work.

29. With reference to Fig. 7, name the type of fuse that should be used in the plug if it is designed to fit into a BS1363 socket?

a BS3871
b BS3036
c BS1362
d BS1361.

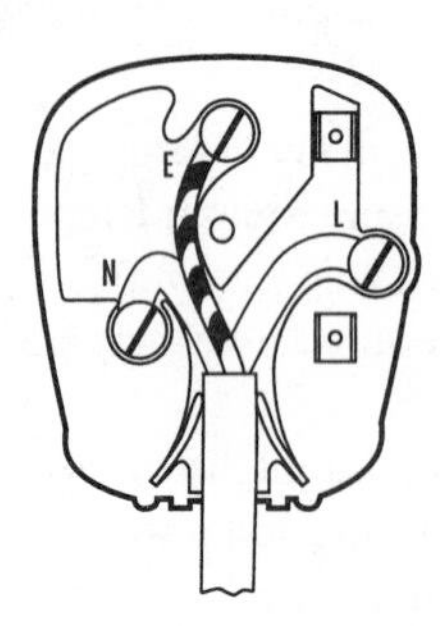

Fig. 7

30. With reference to Question 29, the fuse rating in the plug must *not* exceed

a 15 A
b 13 A
c 10 A
d 3 A.

31. What is the correct name of the clip shown in Fig. 8?

a pipe clip
b earth clip
c bonding clip
d connection clip.

Fig. 8

32. The *IEE Wiring Regulations* are designed to protect persons, property and livestock against

a electric shock, wounds and lightning
b fire, electric shock and burns
c burns, wounds and electric shock
d lightning, burns and injury.

33. Which one of the following is a precautionary measure for working on a 'live' electrical system?

a control of the whole electrical installation
b provision of staff with mixed job experience
c use of emergency escape and first-aid signs
d use of suitable insulated barriers or screens.

34. The term '**dead**' in an electrical sense means

a at or about earth potential and disconnected from any live system
b at a position unlikely to become live if touched by persons
c connected to the neutral point of a three-phase supply system
d taken to ground via an earthed protective conductor.

35. The colour yellow is used to identify BS4343 (BS EN 60309-2) industrial plugs and socket that operate at

a 415 V
b 240 V
c 110 V
d 50 V.

36. Extra-low voltage is a safety voltage that does not exceed

a 120 V ac or 50 V dc
b 120 V dc or 50 V ac
c 110 V dc or 50 V ac
d 110 V ac or 50 V dc.

37. Which one of the following protective devices is required to have a test button?

a miniature circuit breaker
b high-rupturing capacity fuse
c residual current device
d semi-enclosed rewirable fuse.

38. Fig. 9 is a safety sign comprising a white cross on a green background and is used for giving information on/about

a first aid
b dangerous crossings
c emergency escape route
d safe condition.

Fig. 9

39. General rules for observing safe practices on site include being alert, being perceptive, knowing emergency procedures and also

a knowing clocking on/off times
b spending time with other trades
c understanding trade union rules
d protecting oneself and others.

40. In Fig. 10, tilting the head backwards to begin artificial respiration

a allows more blood to circulate around the victim's head
b allows the victim to see the rescuer on recovery
c lifts the victim's tongue off the back of his throat
d provides the rescuer with access to the victim's throat.

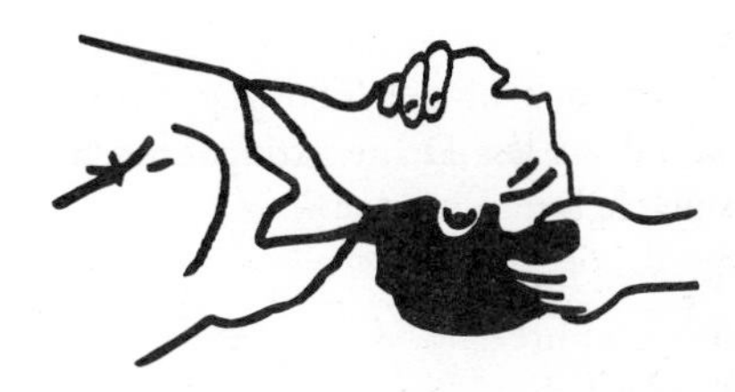

Fig. 10

41. If a working platform on a scaffold is more than 2 m above ground it should be fitted with

a re-inforced steel supports
b flexible expansion joints
c guard rail and toeboard
d non-adjustable pipe clamps.

42. The ladder in Fig. 11 should extend above the landing place by a distance of at least

a 2.5 m
b 2.0 m
c 1.5 m
d 1.0 m.

43. The most probable risk to health whilst carrying an aluminium ladder, are injuries from

a overbalancing and falling down
b touching overhead wires
c severe twisting of the spine
d straining of arms and shoulders.

Fig. 11

44. The rule often used as a guide to the correct angle in which to place and erect a ladder is

a 4:1
b 3:1
c 2:1
d 1:1.

45. Which one of the following types of fire would a Class D fire extinguisher be most suitable for?

a wool fire
b wood fire
c plastic fire
d electric fire.

46. All the following fire extinguishers in Fig. 12 can be used on electrical fires but not

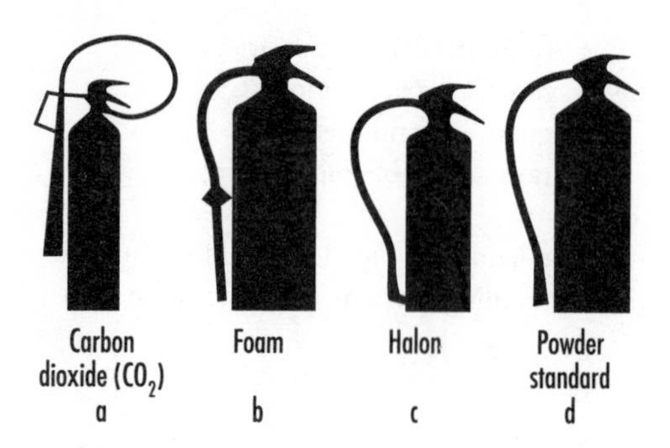

Fig. 12

47. Inflammation of the skin is known as

a dermatitis
b gastritis
c ileitis
d hepatitis.

48. In Fig. 13, guard rails and toe boards are to be provided on the scaffold tower if the

a drop to the ground is more than 1.98 m
b platform is to carry more than three people
c assembly is to be secured to a fixed structure
d castor wheels have no safety locks.

49. Which one of the following is *not* an important consideration when deciding to use a heavy portable hand-held power tool?

a making sure its guard is fitted correctly
b tightening its chuck with the proper key
c using industrial gloves to reduce vibration
d reading the manufacturer's operation instructions.

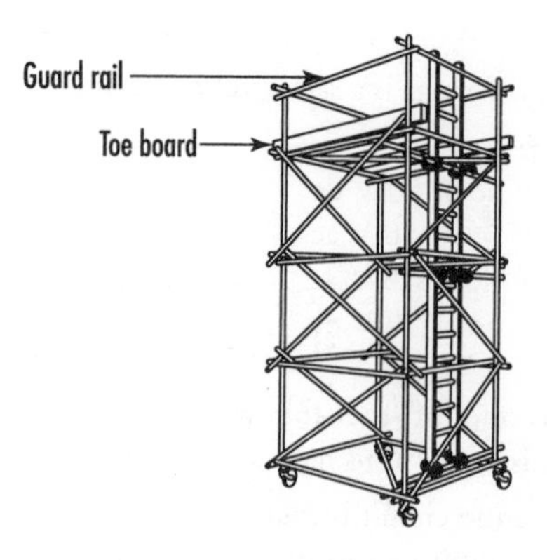

Fig. 13

50. What is the correct name for the access equipment shown in Fig. 14?

a trestle scaffold
b bridge trestle
c wooden platform
d double-step ladders.

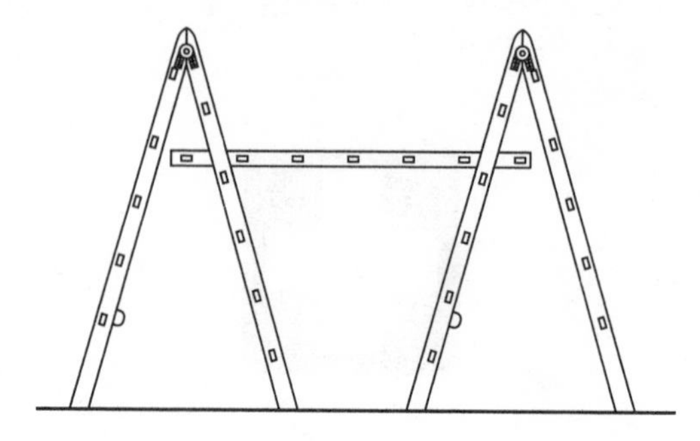

Fig. 14

ASSESSMENT 3.2

Answers, Hints and References

1. **a** *See* Ref. 1, p 13.
2. **b** *See* Ref. 1, p 12.
3. **d** *See* Ref. 4, definitions.
4. **c** *See* Ref. 17, section 7.
5. **a** *See* Ref. 17, section 3.
6. **b** *See* Ref. 1, pp 25–27.
7. **d** *See* Ref. 17, section 3.
8. **d** *See* Ref. 4, definitions.
9. **b** *See* Ref. 4, sections 476 and 537.
10. **a** *See* Ref. 11.
11. **a** *See* Ref. 22.
12. **b** *See* Ref. 11.
13. **c** *See* Ref. 1, p 129.
14. **a** *See* Ref. 17, section 3.
15. **d** *See* Ref. 11.
16. **a**.
17. **b** *See* Ref. 7, p 45.
18. **b** *See* Ref. 1, p 21.
19. **a** *See* Ref. 12, p 91.
20. **c** *See* Ref. 1, p 129.
21. **d**.
22. **d** *See* lamp-crusher manufacturers' information.
23. **c**.
24. **d** *See* Ref. 1, pp 22–23.
25. **b** *See* lamp-crusher manufacturers' information.
26. **a** *See* Ref. 1, p 31.
27. **a** *See* Ref. 11.
28. **a** *See* Ref. 1, pp 31–32. Note words in question.
29. **c** *See* Ref. 19, pp 13–14.
30. **b** *See* Ref. 4, Table 55A.
31. **b** *See* Ref. 1, p 41, Fig. 3.5.
32. **b** *See* Ref. 4, regulation 120–01–01.
33. **d** *See* Ref. 1, p 115 and Ref. 9, regulation 14.
34. **a** *See* Ref. 4, definitions.
35. **c** *See* Ref. 1, p 49.
36. **b** *See* Ref. 1, p 127.
37. **c** *See* Ref. 1, p 22.
38. **a** *See* Ref. 11.
39. **d**.
40. **c** *See* Ref. 19, p 2.
41. **c** *See* Ref. 1, pp 30–31.
42. **d** *See* Ref. 1, p 30, Fig. 2.15.
43. **b** *See* Ref. 1, p 30.
44. **a** *See* Ref. 1, p 30, Fig. 2.15.
45. **d** *See* Ref. 1, p 33.
46. **b** *See* Ref. 1, pp 32–33.
47. **a**.
48. **a** *See* Ref. 1, p30.
49. **c** Gloves will hinder control.
50. **a**.

ASSESSMENT 3.3

1. Which one of the following Acts has the power to stop work immediately by the issuing of a **prohibition notice** on an offending employer?
   a Explosive Act
   b Mines and Quarries Act
   c Health and Safety at Work Act
   d Public Health Act.

2. The *Electricity at Work Regulations* are concerned with risk of death and personal injury from electricity used in
   a high voltage systems only
   b specified installations
   c general work activities
   d mines and quarries only.

3. All the following are employer duties to his employees under the *HSW Act* but *not* the provision
   a and maintenance of plant and systems of work
   b of information, instruction, training and supervision
   c of social entertainment and leisure facilities
   d and maintenance of a working environment that is safe.

4. The *Electricity at Work Regulations 1989* calls for persons to be competent to prevent danger and injury and involve a mix of
   a technical and academic experience
   b installation theory and regulations
   c technical knowledge and experience
   d practical and general knowledge.

5. All the following are functions of a works safety representative, but *not*
   a investigating dangerous occurrences
   b investigating complaints by employees
   c carrying out inspections of premises
   d writing reports for HSE inspectors.

6. If an accident occurs at work which causes incapacity for more than 3 days, it must be reported to the HSE or the
   a local enforcing authority
   b local area hospital board
   c employment service agency
   d public health authority.

7. A formal procedure for a **safe system of work** involves: (i) assessing the task; (ii) identifying any hazards; (iii) defining safe methods; (iv) implementing the system, and also
   a isolating common arrangements
   b monitoring procedures
   c notifying HSE inspectors
   d posting danger notices.

8. The number of safety representatives who can request an employer to set up a safety committee is
   a 5
   b 4
   c 3
   d 2.

9. In the *Electricity at Work Regulations* the term **isolation** means

a removal of the electrical supply from its energy source
b disconnection and separation of electrical equipment
c prohibiting the energy source to be connected to the consumer
d cutting off the electrical supply to switchgear and equipment.

10. In occupied premises, the new emergency safety sign shown in Fig. 1 is to be used

a on all exit routes
b only on non-exit doors
c on every exit door
d in lifts and moving stairs.

Fig. 1

11. In the absence of a trained first-aider, which of the following persons is likely to be asked by an employer to carry out first–aid treatment?

a appointed person
b visiting nurse
c senior apprentice
d part-time worker.

12. Which one of the following substances is least likely to cause industrial dermatitis?

a white spirit
b fibre glass
c vaseline
d tar.

13. Which one of the following hazard symbols, shown in Fig. 2, is used for labelling substances that are very toxic?

a b c d

Fig. 2

14. If a workmate is bleeding from a severe wound, the first step is to

a lie him down and immediately seek help
b clean and wash the injury with a swab
c place a tourniquet above the injury
d apply direct pressure to the injury.

15. In the *IEE Wiring Regulations*, building, utilization and environment are all categories relating to

a external influences
b general characteristics
c periodic inspection
d maintainability.

16. Under the *HSW Act*, all the following are legal duties of an employee to his employer, *except* to

a co-operate in complying with a statutory requirement
b not interfere with or misuse any safety equipment
c take reasonable care to avoid injury to himself and others
d make safe arrangements to transport others to work.

17. In electrical installations, a warning label is required on or adjacent to equipment before gaining access. This requirement is for voltages in excess of

a 500 V
b 415 V
c 350 V
d 250 V.

18. Fig. 3 shows a transformer placed in a position inaccessible to the general public and is a protective measure known as

a isolation from touch
b out of arm's reach
c restrictive location
d indirect contact.

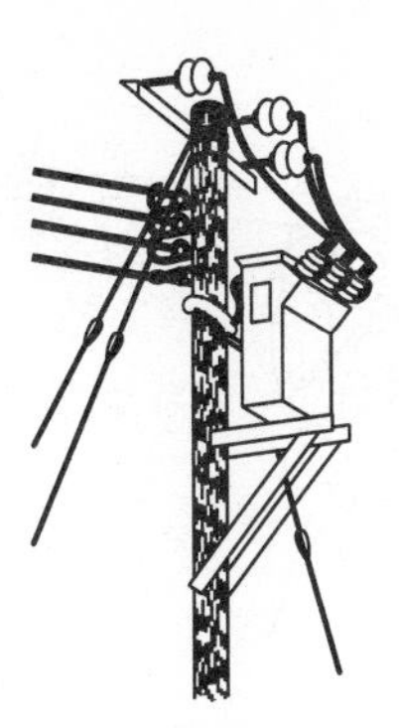

Fig. 3

19. When a person starts on a building site for the first time, he/she should be made aware of the site's
   - **a** safety rules and accident procedures
   - **b** non-nominated subcontractors
   - **c** water and electricity supply points
   - **d** administration and operation policy.

20. In the process of installing circuits, an electrical apprentice accidentally cuts through the insulation of several PVC cables. He should
   - **a** apply additional insulation over the damaged cables
   - **b** rewire the damaged cables if the route is not too long
   - **c** inform the electrician of what damage is done
   - **d** ignore the damage done if it is not too severe.

21. Which one of the following is *least* likely to be the blame for an accident at work to a junior apprentice?
   - **a** basic competence and little knowledge
   - **b** insufficient training and supervision
   - **c** over-enthusiasm and overconfidence
   - **d** improper dress and improper behaviour.

22. Immediate action in the case of a person suffering from the effects of toxic fumes in a closed room is to
   - **a** seek medical help
   - **b** open all windows and doors
   - **c** apply artificial respiration
   - **d** leave the danger area.

23. The statutory instrument *RIDDOR* requires a record to be kept of any injury or disease at work to employees and should include all the following but *not*
   - **a** date, time and place of the event
   - **b** details of persons involved
   - **c** description/nature of the event
   - **d** hours worked by the accident victim.

24. If battery acid is accidentally spilled on a person's skin the affected part should immediately be
   - **a** seen by a general practitioner
   - **b** bandaged with a sterile dressing
   - **c** bathed or flooded with cold water
   - **d** covered with an antiseptic cream.

25. If an electric shock victim does *not* respond to **mouth-to-mouth ventilation** a first-aider should immediately
   - **a** seek medical assistance
   - **b** check for broken limbs
   - **c** keep the victim's body warm
   - **d** check the victim's pulse.

26. A cartridge-operated tool should not be used in an environment that is
   - **a** corrosive
   - **b** humid
   - **c** flammable
   - **d** cold.

27. The electricity safety sign shown in Fig. 4 comes under the category of a
   - **a** prohibition sign
   - **b** warning sign
   - **c** mandatory sign
   - **d** information sign.

Fig. 4

28. Which one of the following is *not* a basic precaution when actually using a portable electric drill?

- **a** checking the fuse and the voltage
- **b** keeping the lead away from the work area
- **c** being aware when the bit pierces the hole
- **d** avoiding additional pressure on the work.

29. Under the *HSW Act* every employer must prepare a written safety policy if he employs

- **a** five or more employees
- **b** four employees
- **c** three employees
- **d** two employees.

30. A permit-to-work form is a safety measure used to protect employees working in

- **a** non-conducting environments
- **b** high earth leakage current areas
- **c** potentially dangerous situations
- **d** restrictive conductive locations.

31. All the following are precautions for working on a live electrical system but *not* necessarily having

- **a** full control of the working area
- **b** use of suitable screens/barriers
- **c** use of suitable test instruments
- **d** employees with mixed experience.

32. With reference to Fig. 5, tightening the terminal screws is best carried by

- **a** holding the plug close to your body
- **b** resting the plug firmly on a flat surface
- **c** resting the plug firmly on your knee
- **d** placing the plug in a clamp or wood vice.

33. With reference to Fig. 5, which one of the following is the *least* important factor while terminating the flexible cable?

- **a** only stripping sufficient insulation from the conductors
- **b** making sure to insert the cable flex into the cable grip
- **c** identifying the internal flex cores
- **d** knowing the type of fuse to be inserted.

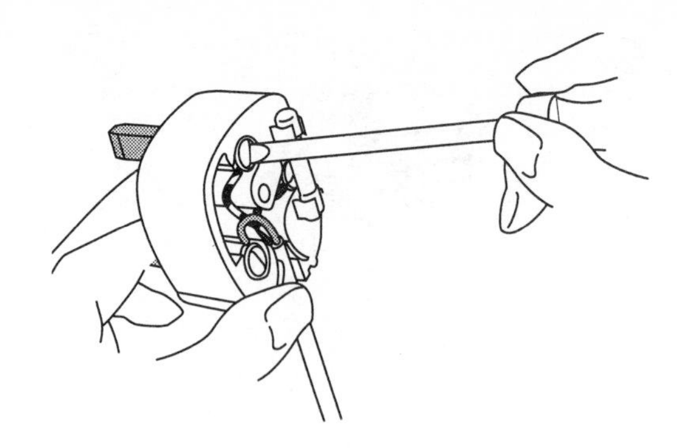

Fig. 5

34. All the following are precautions for an electrical system that is assumed **dead**, *except*

- **a** accepting someone's word that the supply has been isolated
- **b** making sure that all points of the supply are isolated
- **c** securing/locking off each point of isolation
- **d** proving that no supply exists at the point of work.

35. The *IEE Wiring Regulations* are designed to protect persons, property and livestock against the following *except*

- **a** fire
- **b** poison
- **c** electric shock
- **d** burns.

36. In the *Electricity at Work Regulations* the term **isolation** means

- **a** removal of the supply from its energy source
- **b** disconnection and separation of the supply
- **c** prohibiting the energy source to be connected
- **d** not allowing the supply to be connected to equipment.

37. A 240 V BS4343 (BS EN 60309-2) industrial plug and socket (*see* Fig. 6) is identified by the colour

a red
b white
c blue
d yellow.

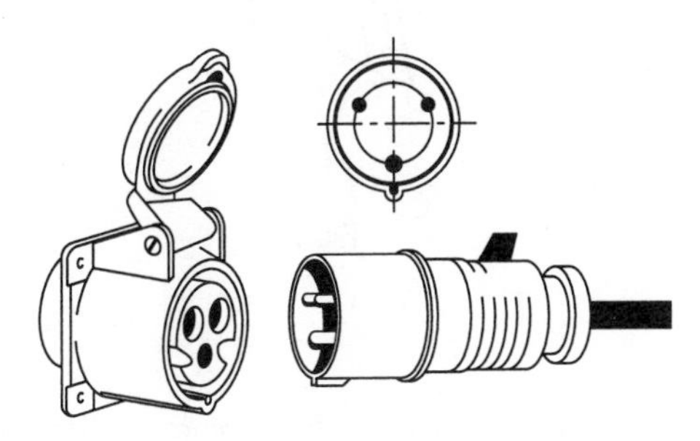

Fig. 6

38. Which one of the following is *not* a method of reducing electric shock?

a insulation or enclosing live conductors
b a low sensitivity residual current device
c double-insulated electrical equipment
d earthing and automatic disconnection of the supply.

39. Electric shock risk can be reduced if the voltage causing it does *not* exceed

a 240 V
b 230 V
c 110 V
d 50 V.

40. A protective device used solely to detect earth faults is called a

a miniature circuit breaker
b high-rupturing capacity fuse
c residual current device
d semi-enclosed rewirable fuse.

41. The minimum room temperature for a comfortable working environment for occupants in a shop or office after the first hour of work is

a 30°C
b 24°C
c 16°C
d 10°C.

42. All the following are places in an electrical installation where you might find a BS951 earth clamp used (*see* Fig. 7), *except*

a on an earthing conductor connected to an earth electrode in soil
b on a main earthing terminal where it is not part of the main switchgear
c on a bonding conductor connected to an extraneous conductive part
d on a protective conductor connected to an earth bar inside a fuseboard.

Fig. 7

43. General rules for observing safe practices on site involve all the following points, *except*

a being alert and perceptive
b knowing emergency procedures
c protecting oneself and others
d awareness of trade union rules.

44. Which one of the following is the *least* possible reasons for evacuating employees from site?

a fire and explosion
b toxic atmosphere
c management meeting
d terrorist activity.

45. The type of fire extinguisher shown in Fig. 8 is used on live circuit fires since it contains

a water
b foam
c dry powder
d CO_2.

Fig. 8

46. A carbon dioxide fire extinguisher

a knocks flames down with shear pressure
b smothers flames by displacing oxygen in the air
c cools flames with its high composition of water
d starves flames with its multichemical mixture.

47. One method of securing the foot of a ladder is to use

a a stake driven into the ground close to the base
b the hook of a long metal crowbar anchored to the ground
c wheelbarrow handles through the lower rungs
d several building bricks placed at the base.

48. With reference to Fig. 9, what is the name given to the two side members of a ladder?

a runners
b stiles
c rails
d limbs.

Fig. 9

49. In Fig. 10 what is the recommended safe climbing angle?

a 85°
b 75°
c 65°
d 55°.

Fig. 10

50. If a working platform on a scaffold is more than 2 m above ground, it should be fitted with

a reinforced steel supports
b flexible expansion joints
c guard rail and toeboard
d non-adjustable pipe clamps.

ASSESSMENT 3.3

Answers, Hints and References

1. **c** *See* Ref. 1, p 12.
2. **c** *See* Ref. 1, p 13.
3. **c** *See* Ref. 1, pp 12–13.
4. **c** *See* Ref. 1, pp 113–115.
5. **d** *See* Ref. 1, pp 13–15.
6. **a** *See* Ref. 1, p 15.
7. **b** *See* Ref. 1, pp 25–27.
8. **d** *See* Ref. 1, p 14.
9. **b** *See* Ref. 1, pp 113–115.
10. **a** *See* Ref. 11.
11. **a** *See* Ref. 1, pp 20–21.
12. **c** *See* Ref. 15.
13. **a** *See* Ref. 11.
14. **d** *See* Ref. 1, p 21.
15. **a** *See* Ref. 4, Appendix 5.
16. **d** *See* Ref. 1, pp 12–13.
17. **d** *See* Ref. 4, regulation 514–10–01.
18. **b** *See* Ref. 4, regulation 412–05–02.
19. **a** *See* Ref. 17, employee duties.
20. **c** This is the person supervising training.
21. **a** It is the least possible reason.
22. **d**.
23. **d** *See* Ref. 1, pp 116–120.
24. **c**.
25. **d** *See* Ref. 1, p 23.
26. **c** *See* Ref. 19, p 100.
27. **a** *See* Ref. 11.
28. **a** *See* Ref. 1, p 31.
29. **a** *See* Ref. 1, p 13.
30. **c** *See* Ref. 1, p 131, answer 3.
31. **d** *See* Ref. 10, regulation 14 and Ref. 1, p 115.
32. **b** The screwdriver could easily slip and cause injury.
33. **d** The question refers to actually terminating.
34. **a** *See* Ref. 10, regulation 13 and Ref. 1, p 114.
35. **b** *See* Ref. 4, regulation 120–01–01.
36. **b** *See* Ref. 10, regulation 12 and Ref. 1, p 114.
37. **c** *See* Ref. 1, p 49, Fig. 3.15.
38. **b** *See* Ref. 1, p 22.
39. **d** *See* Ref. 1, pp 22–23, Fig. 2.7.
40. **c** *See* Ref. 1, p 22, Fig. 2.6.
41. **c** *See* Ref. 23.
42. **d** *See* Ref. 4, regulation 514–13–01.
43. **d** It does not directly involve safety.
44. **c**.
45. **d** *See* Ref. 1, p 33.
46. **b** *See* Ref. 1, p 33.
47. **a** *See* Ref. 1, p 30.
48. **b** *See* Ref. 19, p 7, Fig. 1.7.
49. **b** *See* Ref. 1, p 30.
50. **c** *See* Ref. 1, p 30, Fig. 2.15.

4

PROCEDURES AND PRACTICES FOR MOVING LOADS

To tackle the assessments in Topic 4 you will need to know:

- numerous definitions associated with the movement of loads;
- safety requirements concerning the manual handling, lifting and movement of loads;
- methods of lifting loads that are operated by different power sources;
- methods of transporting loads such as flat trailers, hand trucks and fork-lift trucks;
- recognized safety signs used for warning personnel at work that loads are being moved.

DEFINITIONS

Block and tackle – a rope and pulley system used for lifting loads vertically by pulling down on the free end of the rope.

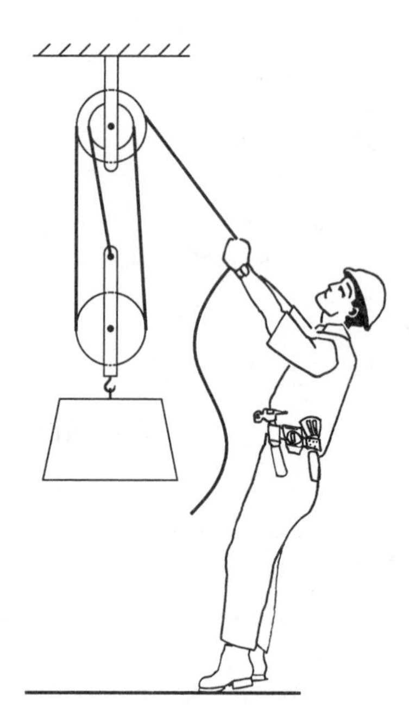

Block and tackle being used to lift a heavy load

Centre of gravity – a point through which the whole weight of an object seems to act.
Chocking – a method using wooden wedges to stop pipes and drums from rolling, particularly those stored on top of each other.
Efficiency – of a machine, is the ratio of **work got out** (work done on the load) and **work put in** (work done by effort).
Effort – work or energy that is put in or a force applied to an object by a machine to move it.
Fulcrum – a point of support on which a lever pivots.
Lifting sources – methods commonly used to lift loads, such as **manual power** (e.g. arms), **pneumatic power** (operated by compressed air), **mechanical power** (operated by gear trains, winches etc), **electric power** (operated by a motor), **hydraulic power** (operated by pressure from a fluid).

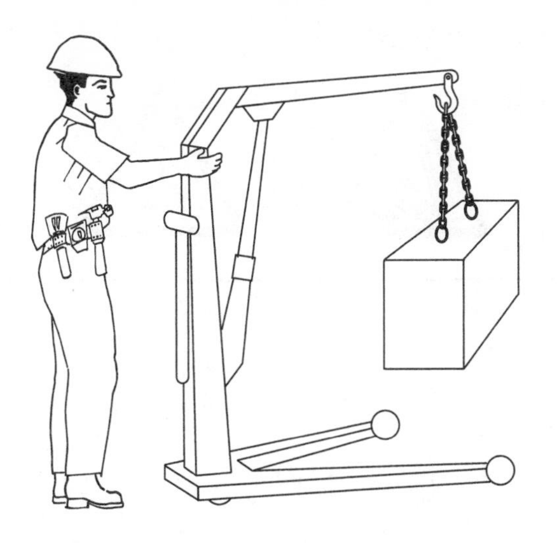
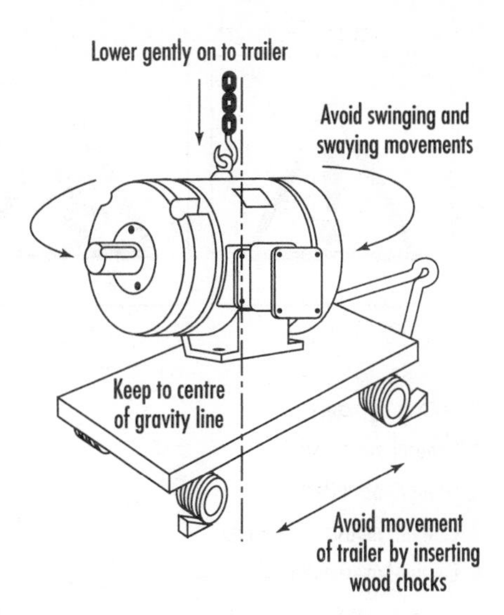

Hand-powered hydraulic lift being used to move a heavy load

Load – a discrete movable object, e.g. something being carried. It can also describe a force exerted by a machine.
Manual handling operations – the movement of a load(s) by human effort
Mechanical advantage – a ratio expressing load/effort, used to tell how much a lever is able to magnify a force.
Null point – a void occurring in a magnetic field where no lines of force are present, often referred to as a neutral point.
Neutral equilibrium – a term referring to an object whose centre of gravity does not rise or fall if it is moved, e.g. a billiard ball.
Personal protective equipment – clothing such as gloves, aprons, safety shoes etc recommended for the task.

Adopt the correct procedure when lifting a heavy load

Safe working load – the mass in kilograms that lifting equipment can safely lift
Torque – a force as a result of a turning effort or turning moment. Motors develop starting torques, accelerating torques and running torques.
Velocity ratio – a ratio expressing: distance moved by the effort/distance moved by the load (in the case of a lever) or number of teeth on a driven wheel/number of teeth on a driving wheel (in the case of a gear train). For pulley systems it is determined by the number of pulleys used or number of ropes supporting the load.

1

2

3

4

Warning symbols

1 General danger sign

2 Caution, obstacles

3 Caution, industrial vehicles

4 Caution, overhead load

TOPIC 4

Procedures and Practices for Moving Loads

ASSESSMENTS 4.1 – 4.3

Time allowed: 1 hour

Instructions

* You should have the following:

  Question Paper
  Answer Sheet
  HB pencil
  Metric ruler

* Enter your name and date at the top of the Answer Sheet.
* When you have decided a correct response to a question, on the Answer Sheet, draw a straight line across the appropriate letter using your HB pencil and ruler (see example below).
* If you make a mistake with your answer, change the original line into a cross and then repeat the previous instruction. There is only one answer to each question.
* Do not write on any page of the Question Paper.
* Make sure you read each question carefully and try to answer all the questions in the allotted time.

Example:

a 400 V	**a** 400 V
~~**b**~~ 315 V	~~**b**~~ (crossed) 315 V
c 230 V	**c** 230 V
d 110 V	~~**d**~~ 110 V

ASSESSMENT 4.1

1. The *Manual Handling Operations Regulations 1992* define a load as a discrete movable object such as a

   a sledge hammer used to break rocks
   b quantity of sand carried on a shovel
   c hoist used to carry bricks on a building site
   d wheelbarrow used to move building material.

2. With reference to Fig. 1, to avoid injury, you should

   a allow your elbows to freely bend
   b lift the load using your leg muscles
   c keep both feet flat on the ground
   d quickly jerk the load towards your chest.

Fig. 1

3. Which one of the following is *not* an example of good posture when lifting a load?

   a keep the spine straight
   b keep the shoulders level
   c keep feet slightly apart
   d keep heels off the ground.

4. All reasonable heavy loads that are manually handled should only be lifted

   a if left in a dangerous place
   b after obtaining written consent
   c if no assistance is available
   d after making proper assessment.

5. With reference to Fig. 2, which one of the following is *not* an important consideration?

   a both persons should be about the same height
   b one person should give the instructions
   c both persons should share equal mass
   d the load should be kept at shoulder height.

Fig. 2

6. The point on a ladder to achieve balance so that it can be comfortably carried is called the

a zero mass position
b centre of gravity
c neutral plane
d reference position.

7. To lift correctly the load on the table in Fig. 3, the person should firstly

a kneel on the floor obstruction
b reach over the floor obstruction
c remove the floor obstruction
d stand on the floor obstruction.

Fig. 3

8. Fig. 4 represents different shaped objects resting on a flat surface. Which one is the *least* stable?

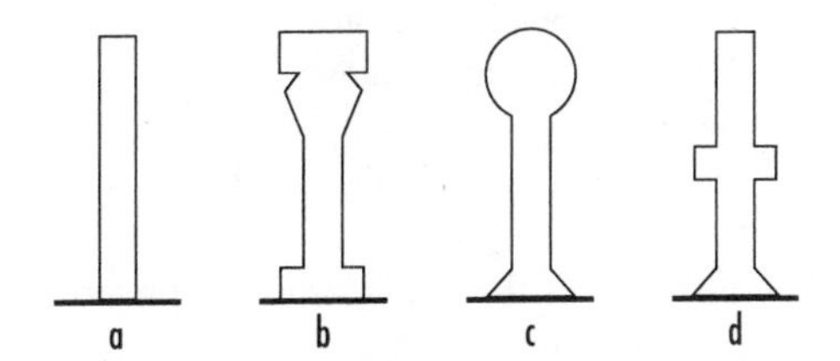

Fig. 4

9. The abbreviation **SWL** is often stated on lifting equipment and means

a steady weight limit
b safe working load
c stable working limit
d short wide load.

10. Which one of the following is *not* a type of power lifting source?

a hydraulic power
b pneumatic power
c nuclear power
d muscle power.

11. It is generally recommended that power lifting equipment be used for loads in excess of

a 100 kg
b 80 kg
c 50 kg
d 20 kg.

12. What is the name of the lifting apparatus shown in Fig. 5?

a block and tackle
b wheel and axle
c rope winch
d pulley hoist.

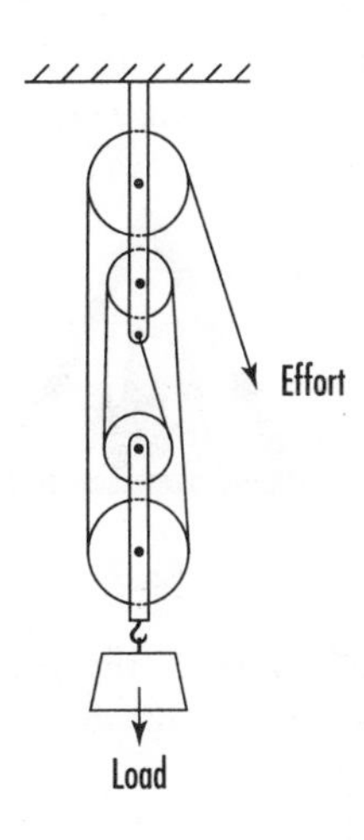

Fig. 5

13. In Fig. 5, **load** divided by **effort** is called

a velocity ratio
b mechanical advantage
c efficiency
d torque.

14. The effect of the idle wheel marked *C* in Fig. 6, is to cause gear wheels *A* and *B* to move

- **a** in the same direction
- **b** in the opposite direction
- **c** at half their speed
- **d** at different speeds.

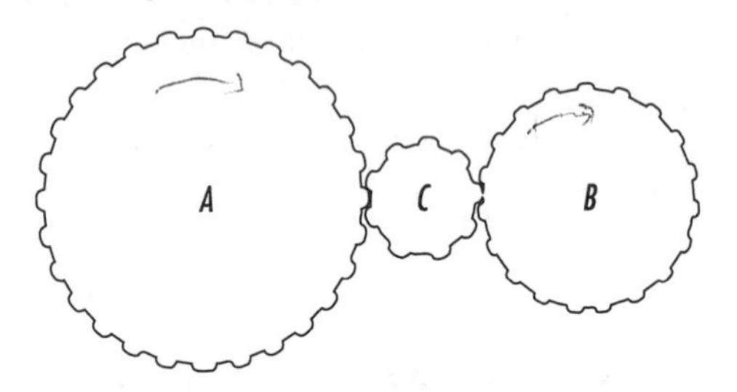

Fig. 6

15. The warning sign shown in Fig. 7 (black on yellow background) is most likely to mean

- **a** vehicle access area
- **b** caution, industrial vehicles
- **c** danger, vehicles crossing
- **d** vehicle loading bay area.

Fig. 7

16. In Fig. 8, what is the protective measure called which prevents the pipes from rolling?

- **a** franking
- **b** shanking
- **c** blocking
- **d** chocking.

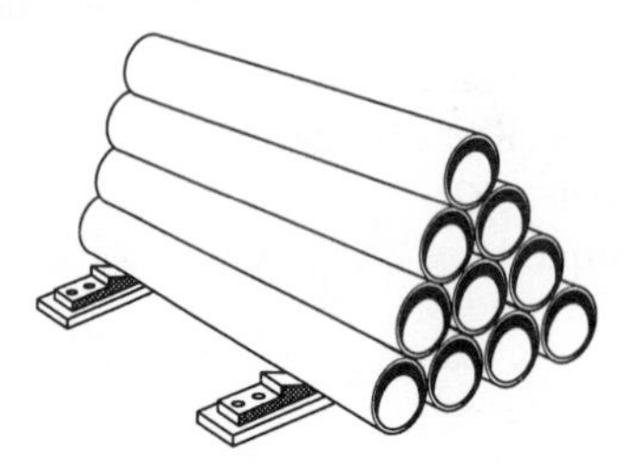

Fig. 8

17. In the process of moving a load with a crane, positioning the hook above the **centre of gravity** of the load

- **a** allows equal sling length
- **b** creates a safe work height
- **c** provides an even balance
- **d** reduces sling tension.

18. When a nut is being tightened with a spanner, the turning moment is often called the

- **a** torque
- **b** leverage
- **c** effort
- **d** tension.

19. A pneumatic lifting machine is one that is filled with or operated by

- **a** transformer oil
- **b** liquid nitrogen
- **c** compressed air
- **d** hydrogen gas.

20. When carrying a long ladder on your shoulder, you should make sure that the end behind you is

- **a** higher than the front end
- **b** kept reasonably low
- **c** displaying a red flag
- **d** not over 4 m long.

21. One should avoid pushing or pulling a heavy load supported by a crane's sling as this might cause the load's

- **a** supports possibly to break
- **b** centre of gravity to move
- **c** weight to fluctuate
- **d** contents to become damaged.

22. Which one of the following is the preferred method of transporting a heavy motor over a flat, unobstructed concrete floor?

- **a** four-wheel flat trailer
- **b** builders' wheelbarrow
- **c** low-tech sack trolley
- **d** block and tackle.

23. All the following are good reasons for splitting loads into smaller weight categories for handling purposes, *except*

a items are easier to grasp and move
b items are less likely to become damaged
c items can be individually identified
d items can be stored more easily.

24. The abbreviation **PPE** is often used in safety regulations to mean

a personal protective equipment
b private practice engineering
c personal property enquiries
d peak performance entries.

25. Which one of the following is *not* an advantage of using a hoist on a building site?

a large loads can be moved vertically
b loads can be taken to different levels
c labour-time is saved when working above ground
d manual handling accidents are reduced.

ASSESSMENT 4.1

Answers, Hints and References

1. **b** *See* Ref. 24, regulation 2, Clause 14.
2. **b** *See* Ref. 1, pp 28–29.
3. **d** *See* Ref. 24, clauses 54 and 163.
4. **d** *See* Ref. 24, clause 33.
5. **d**.
6. **b** *See* Ref. 6, pp 18–19.
7. **c**.
8. **b** *See* Ref. 6, p 19.
9. **b** *See* Ref. 1, p 29.
10. **c**.
11. **d** *See* Ref. 24, clause 9 and Appendix 1.
12. **a**.
13. **b** *See* Ref. 6, p 27.
14. **a**.
15. **b** *See* Ref. 10.
16. **d** *See* Ref. 1, pp 28–30.
17. **c** *See* Ref. 1, p 29.
18. **a** *See* Ref. 6, p 20.
19. **c**.
20. **b** *See* Ref. 1, p 30.
21. **b** *See* Ref. 6, p 19.
22. **a** *See* Ref. 1, p 30, Fig. 2.14.
23. **d** *See* Ref. 24, regulation 4, clauses 136–143.
24. **a** *See* Ref. 24, definition and regulation 4, clause 103.
25. **d**.

ASSESSMENT 4.2

1. Which one of the following is regarded as a load, as defined in the *Manual Handling Operations (MHO) Regulations 1992*?
   - **a** a patient being moved onto a hospital bed
   - **b** a ladder being used to climb a wall
   - **c** a chainsaw cutting through a tree
   - **d** a crowbar being used to lift a motor.

2. In the *MHO Regulations*, the term **manual handling** refers to all of the following *except*
   - **a** transporting a load
   - **b** supporting a load
   - **c** lashing down a load
   - **d** throwing down a load.

3. With reference to Fig. 1, which one of the following could prove to be a hindrance?
   - **a** tilting the drum before rolling it
   - **b** keeping correct balance and posture
   - **c** wearing industrial protective gloves
   - **d** wearing industrial protective shoes.

Fig. 1

4. For a load to be more stable its **centre of gravity** should be
   - **a** low and it should have a wide base
   - **b** acting equally across all surfaces
   - **c** in the middle of its mass
   - **d** at a point approaching zero.

5. The guideline weight given in the *MHO Regulations* for holding a load at arm's length while standing in a stable position is
   - **a** 50 kg
   - **b** 25 kg
   - **c** 15 kg
   - **d** 10 kg.

6. If the round ball in Fig. 2, is placed on a level base and is stationary, it is said to have
   - **a** neutral equilibrium
   - **b** potential force
   - **c** zero stability
   - **d** null balance.

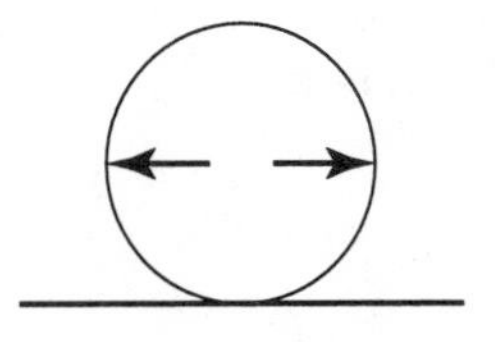

Fig. 2

7. All the following are important considerations before attempting to manually lift a load, *except*

   a making trial lifts on similar loads
   b getting help if load is too heavy
   c checking the load for any sharp edges
   d checking the load for adequate handling space.

8. What type of power source operates the small hoist shown in Fig. 3?

   a pneumatic power
   b hydraulic power
   c magnetic power
   d electric power.

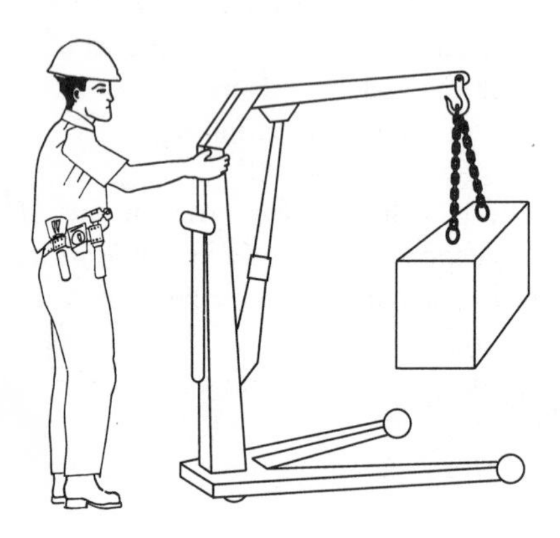

Fig. 3

9. Which one the following is *not* a suitable method of moving the large motor in Fig. 4 some distance?

   a steel rollers
   b coconut mat
   c block and tackle
   d crowbars.

Fig. 4

10. At which point in Fig. 5 is **effort** being applied to lift the load?

   a W
   b X
   c Y
   d Z.

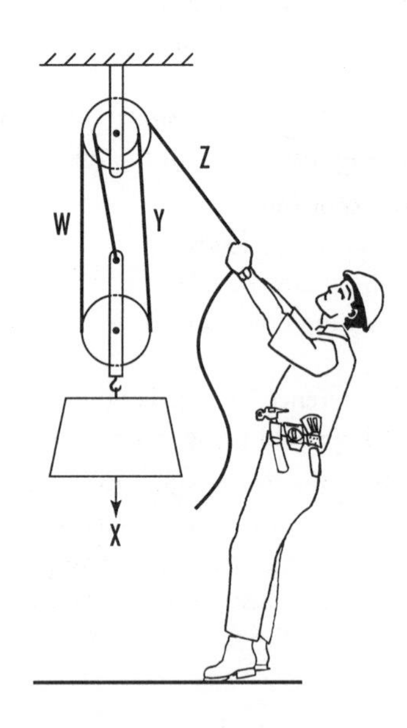

Fig. 5

11. In Fig. 6, the sling is placed above the **centre of gravity** of the load so as to

a maintain a balance
b avoid twisting movements
c avoid mechanical stress
d position the trailer.

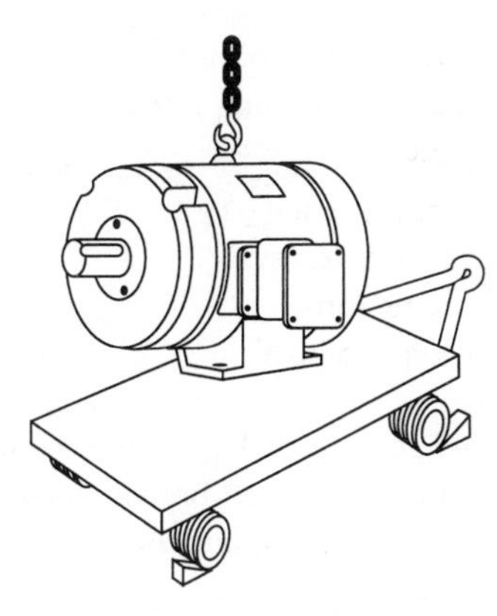

Fig. 6

12. All the following are important considerations when using armoured cable jacks, *except*

a being aware of any sudden collapse
b keeping the drum evenly balanced
c making sure the ground is firm
d keeping the support bar greased.

13. The point on which a lever pivots is called the

a middle
b fulcrum
c hinge
d centre.

14. The safety warning sign shown in Fig. 7 is used where there is a

a crane in constant use
b load passing overhead
c mobile crane in use
d need for safety hats.

Fig. 7

15. Which one of the following is *not* a method of moving a load relying upon the use of manual power?

a screwjack
b pulley block
c winch
d electromagnet.

16. What is the name of the force that is produced by the earth's gravity?

a weight
b tensile
c pressure
d torque.

17. The type of load lifting device shown in Fig. 8 is called a

a wheel and ratchet
b wheel and axle
c cable drum jack
d hand-winch.

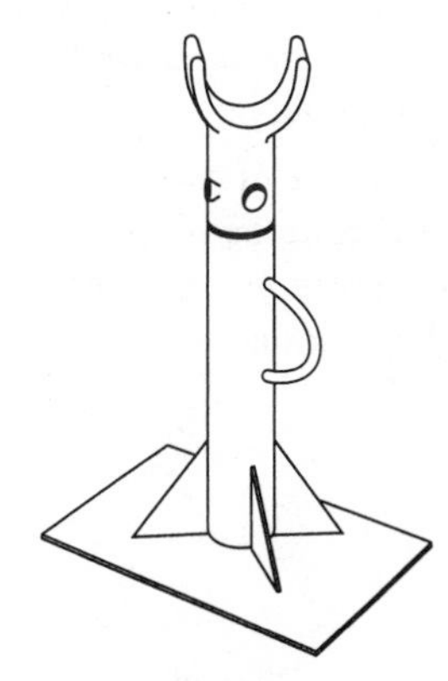

Fig. 8

18. Where loads are being moved by an overhead crane, all the following should be avoided, *except*

a people underneath the load
b leaving the load suspended
c allowing the load to swing
d keeping a safe working load.

19. To avoid a load accidentally moving before any lifting equipment is removed, it should be made

a rigid
b sound
c secure
d tethered.

20. Which one of the following load moving machines operates by using a fluid?

a gantry crane
b hydraulic hoist
c power vacuum lifter
d screwjack.

21. All of the following are safety measures likely to reduce back injury during the manual moving of heavy loads, *except*

a keeping the load close to the body
b controlled pushing and pulling
c provision of compulsory rest pauses
d job rotation using different muscles.

22. When a reasonably heavy load is lowered to the ground by hand, to avoid injury to one's fingers, it should be

a tilted on one of its corners
b placed on strong supports
c gently rested on one knee
d held by a strong rope.

23. What is the preferred method of transporting 200 kg of loose sand over rough ground?

a flat trailer
b large bucket
c sack truck
d wheelbarrow.

24. The main reason for not wearing gloves during manual handling operations is because they could

a cause blistering
b impair dexterity
c become torn
d become trapped.

25. With reference to Fig. 9, the manual handling of the motor from the lorry requires not only teamwork but also

a intuition and motivation
b skill and experience
c ability and qualifications
d knowledge and understanding.

Fig. 9

ASSESSMENT 4.2

Answers, Hints and References

1. **a** *See* Ref. 24, regulation 2, clause 14.
2. **c** *See* Ref. 24, regulation 2, clause 16.
3. **c** *See* Ref. 24, regulation 4, clause 103.
4. **a** *See* Ref. 6, p 19.
5. **d** *See* Ref. 24, Appendix 1.
6. **a** *See* Ref. 6, p 19.
7. **a** It would defeat the object of safety.
8. **b**.
9. **d** *See* Ref. 1, p 29, Fig. 2.12(c).
10. **d** *See* Ref. 6, p 28, Fig. 2.20.
11. **a** *See* Ref. 6, p 19.
12. **d**.
13. **b**.
14. **b** *See* Ref. 11.
15. **d**.
16. **a** *See* Ref. 6, p 15.
17. **c**.
18. **d** *See* Ref. 1, p 29.
19. **c**.
20. **b**.
21. **c** *See* Ref. 24, regulation 4, clause 125.
22. **b**.
23. **d**.
24. **b** *See* Ref. 24, regulation 4, clause 103.
25. **b**.

ASSESSMENT 4.3

1. The *Manual Handling Operations (MHO) Regulations 1992* provide
   - **a** general notes for employees who have to lift and move heavy loads
   - **b** statutory requirements for all people who have to lift and move loads
   - **c** specifications for loads that need lifting and moving
   - **d** statutory rules for all load-lifting apparatus and equipment.

2. Which one of the following is *not* a musculoskeletal disorder?
   - **a** bruised joint
   - **b** swelling tendon
   - **c** mental stress
   - **d** back ache.

3. All of the following are risk assessment areas that employers need to consider to satisfy the *MHO Regulations*, *except* the assessment of
   - **a** both task and load
   - **b** any lifting apparatus
   - **c** the working environment
   - **d** an individual's capability.

4. The *MHO Regulations* provide a guideline weight for the person carrying the load in Fig. 1 which is between
   - **a** 25 and 50 kg
   - **b** 20 and 25 kg
   - **c** 10 and 20 kg
   - **d** 5 and 10 kg.

5. When the person in Fig. 1 reaches the delivery point and lowers the load to the floor, it would be advisable for him to
   - **a** slightly twist his body on one side
   - **b** bend forward keeping both legs straight
   - **c** keep his back and head slightly bent
   - **d** pause and rest his elbows on his thighs.

Fig. 1

6. The point through which the whole weight of an object seems to act is called the
   - **a** fulcrum
   - **b** load centre
   - **c** centre of gravity
   - **d** neutral zone.

7. Name the bone labelled *X* in Fig. 2 which has to bear the weight of the whole of the body lying above it.

   a dorsal bone
   b coccyx bone
   c lumbar bone
   d sacrum bone.

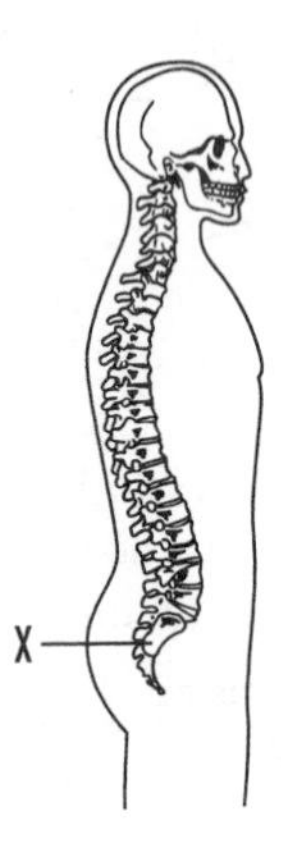

Fig. 2

8. All of the following are groups of bones associated with the spinal column *except* the

   a skull
   b neck
   c back
   d waist.

9. To avoid the need for employees to undertake manual handling operations that might involve a risk of injury, it is a duty of all employers to initially

   a make an assessment of the work
   b engage specialist work gangs
   c provide adequate lifting apparatus
   d issue protective equipment.

10. Which one of the following signs shown in Fig. 3 is used to warn against general danger?

Fig. 3

11. One method of team-handling the motor shown in Fig. 4 is to move it by using a

   a sack trolley
   b coconut mat
   c steel frame glider
   d sloping wooden plank.

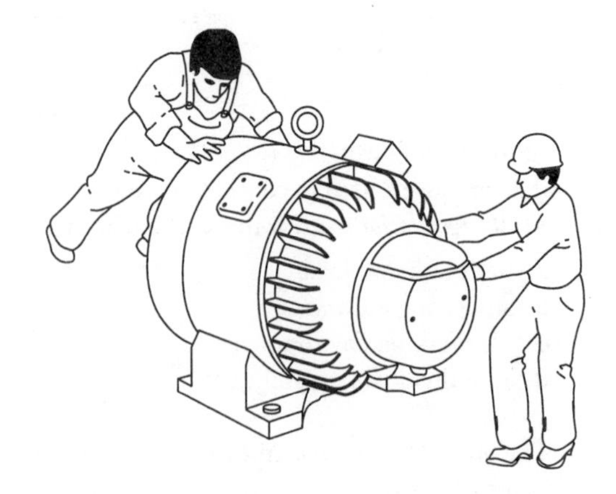

Fig. 4

12. The *MHO Regulations* state that an approximate guide to the capability of a manual handling team consisting or two persons should be

a three-quarters the sum of their individual capabilities
b two-thirds the sum of their individual capabilities
c one-third the sum of their individual capabilities
d one-quarter the sum of their individual capabilities.

13. The person in Fig. 5 is lifting the load using a **mechanical power source** called a

a fixed pulley system
b single rope hoist
c wheel and axle
d block and tackle.

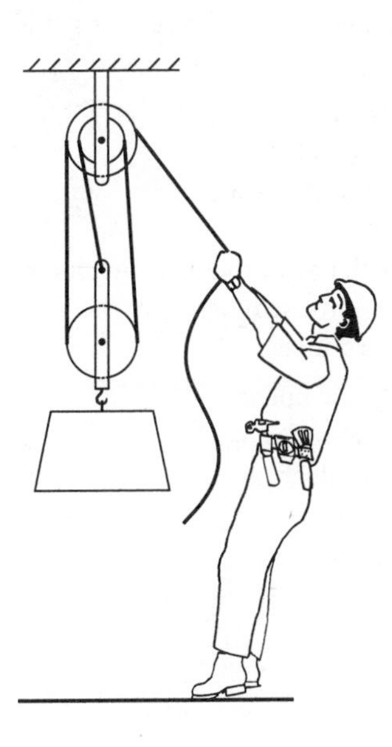

Fig. 5

14. The term **mechanical advantage** is used to describe how much a force can be magnified and is found by the ratio:

a load/effort
b output/input
c power/time
d torque/distance.

15. In Fig. 6, when the smaller driving gear rotates one complete revolution, the larger driven gear rotates

a ¾ of a revolution
b ⅔ of a revolution
c ½ of a revolution
d ¼ of a revolution.

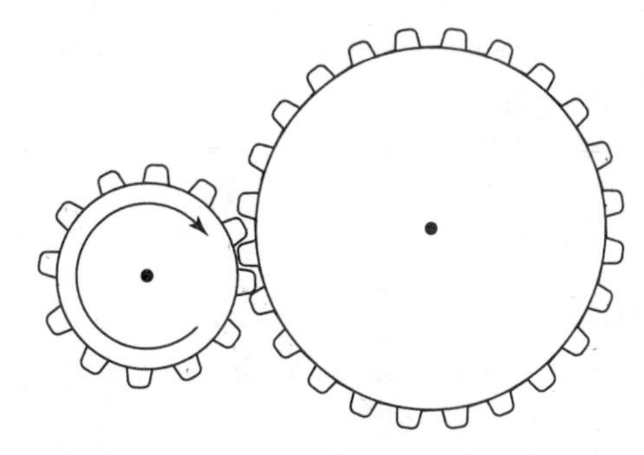

Fig. 6

16. With reference to Question 15 above the **velocity ratio** is

a 4
b 3
c 2
d 1.

17. The wheelbarrow shown in Fig. 7 is a Type 2 lever with the load force acting in the direction labelled

a *W*
b *X*
c *Y*
d *Z*.

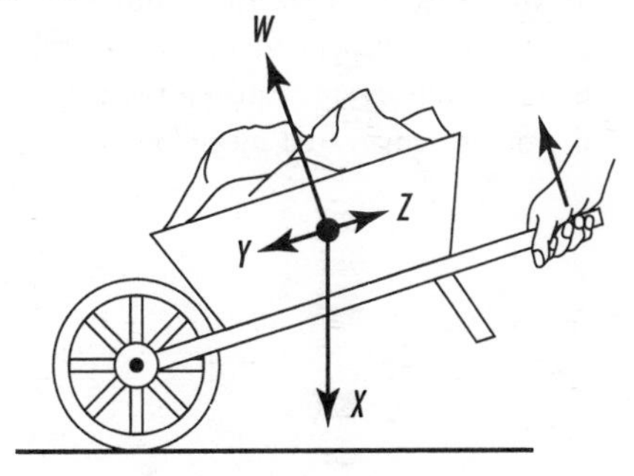

Fig. 7

18. The power source associated with a diesel engine is

a mechanical
b pneumatic
c hydraulic
d combustion.

19. The term 'shock loading' is often associated with lifting equipment that is allowed to

a support loads too long
b rapidly swing or twist
c be operated incorrectly
d operate inefficiently.

20. One advantage of using a lever is that it

a has a very small mechanical advantage (i.e. load/effort)
b saves purchasing other expensive lifting gear
c reduces the force required to move the load
d extends the centre of gravity of the load being moved.

21. When lifting a load with a chain and hook, you should position the hook

a above the centre of gravity
b where least damage will occur
c as close as possible to the fulcrum
d at the easiest support point.

22. All of the following are to be avoided when transporting a load with a crane, *except*

a areas where people are working
b loads left suspended and unsupervised
c jerky movements causing twisting
d routes demarcated by yellow lines.

23. A safety precaution that needs to be considered during the off-loading of the motor in Fig. 8 is to

a tie a red warning flag on the lorry
b secure the load with two guide ropes
c place supports under the wooden planks
d avoid too many handlers on the lorry.

Fig. 8

24. Which one of the following load transporting methods can be very unstable?

a sack trolley
b flat trailer
c wheelbarrow
d fork lift truck.

25. Which one of the following levers has its fulcrum situated between the effort and the applied load?

a bottle opener
b coal-tongs
c wheelbarrow
d pliers.

ASSESSMENT 4.3

Answers, Hints and References

1. **b** *See* Ref. 24, regulation 2, clause 14.
2. **c** *See* Ref. 25.
3. **b** *See* Ref. 24, regulation 4, clause 34.
4. **b** *See* Ref. 24, Appendix 1.
5. **d** *See* Ref. 24, clauses 153–164.
6. **c** *See* Ref. 6, p 19.
7. **d**.
8. **a**.
9. **a** *See* Ref. 24, clause 35.
10. **a** *See* symbols on the definition page.
11. **b** *See* Ref. 1, pp 28–30, Fig. 2.13.
12. **b** *See* Ref. 24, regulation 4, clause 72.
13. **d** *See* Ref. 6, p 28.
14. **a** *See* Ref. 6, p 27.
15. **c** Count the teeth of the gears (12/24).
16. **c** *See* velocity ratio on the definition page.
17. **b** *See* Ref. 6, p 21.
18. **d**.
19. **b**.
20. **c** *See* Ref. 6, pp 20–21.
21. **a** *See* Ref. 6, p 19.
22. **d**.
23. **c**.
24. **c**.
25. **d** *See* Ref. 6, pp 20–21.

5

PROCEDURES AND PRACTICES FOR WORK PREPARATION AND COMPLETION

To tackle the assessments in Topic 5 you will need to know:

- numerous definitions associated with the procedures and practices for work preparation and completion;
- how the use measuring and setting out equipment such as rulers, tapes, callipers, spirit levels and water levels;
- how to take in situ measurements for fabricating conduit, trunking and traywork;
- block, circuit, wiring, schematic and layout diagrams and be able to transfer information from drawings to the installation site;
- procedures for recording receipts and checking materials and tools as well as the procedure for security of tools and equipment;
- how to complete time sheets, job sheets and daywork sheets as well as write reports;
- how to protect electrical work and equipment during the installation process and also be able to identify suitable fixing methods for general and specific applications;
- how to access arrangements for locating wiring systems and equipment during the erection, completion and use of the installation;
- suitable methods of restoring the fabric of a building, affected by electrical work, and also the procedures for the removing and disposing of waste materials and substances.

DEFINITIONS

Block plan – a plan that identifies a site or locates the outline of proposed buildings in relation to a wider area.
Bolster chisel – a steel chisel similar to a cold chisel but having a wide wedge-shaped end, used for lifting floorboard or chasing walls.
Calliper – a measuring tool used for measuring the outside or the inside diameter of tubular or hollow-shaped objects, such as pipes. It comprises two legs lightly riveted together at one end so as to form a pivot while the open legs span the distance to be measured.

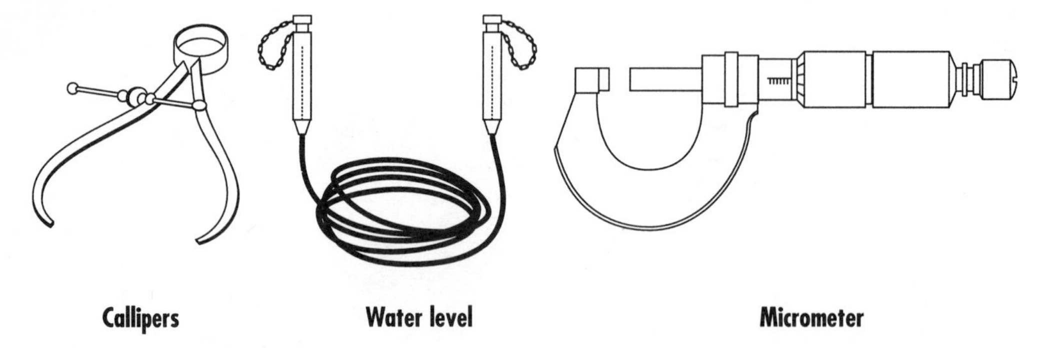

Callipers **Water level** **Micrometer**

Centre punch – a marking out tool made of hard carbon steel used for making small indentations on metal surfaces.
Chalk line – a line coated with chalk that is strained between two predetermined points and plucked by hand to snap against a surface leaving on the surface a line of chalk.
Cold chisel – a hexagonal or octagonal shaped tool made from carbon steel used for cutting off rough metal, rivets or bolts. It is tapered at one end to form a fine hard cutting edge.
Compasses – a steel, sharp pointed marking out instrument used for scribing circles and sometimes curves on a metal surface.
Conduit – a hollow metal or plastic pipe made in various diameters and of standard length, used for the mechanical protection of insulated cables.
Corrosion inhibitor – a method of treatment against corrosion by coating the surface of a metal or removing its corrosive element.
Daywork sheet – the recording of all time and material for unavoidable work outside the scope of a contract.

Die – part of a tool that is used for making an external thread on the end of a conduit or piece of tubing.
Electrical diagrams – block diagrams show only one line connection between major items of equipment. Circuit diagrams show the actual circuit conductors and connections between major items of equipment. Wiring diagrams are similar to circuit diagrams but provide more details about connections. Schematic diagrams are single-line diagrams used to simplify complicated circuits by showing only the essential operating/control circuitry. Layout diagrams are often plans showing where electrical items are positioned.

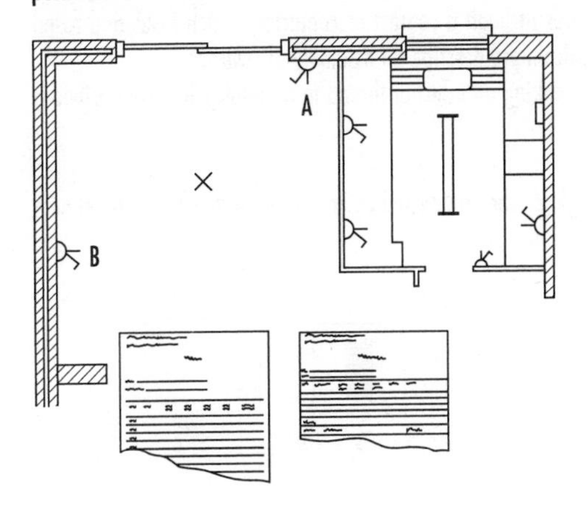

Layout drawing

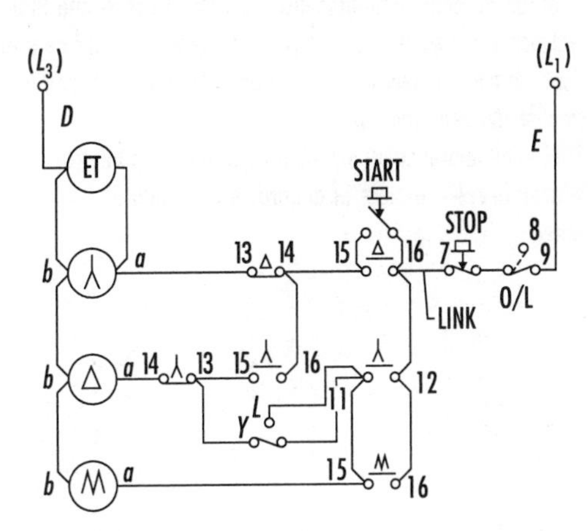

Schematic diagram

First-fix stage – a building operation stage often after weather-proofing whereby the contractor/subcontractors are able to start their internal installations (e.g. plumbers commence pipework and tanks; electricians erect conduit and wiring; plasterers start ceilings, walls, rendering etc).
Fishplate – a flat piece of metal shaped to secure a join, often in steel trunking, during fabrication work.
Job sheet – an instructional programme of work that is required to be done.
Micrometer – an adjustable instrument that is extremely accurate for measuring small lengths and angles.
OHLS – a wiring system abbreviation meaning zero halogen low smoke.
Philblock – a fixing designed for casting in concrete comprising a tough composition moulded block.
Plumb line – a weighted plumb bob attached to a line used for finding the vertical when marking-off walls etc.
Rawlbolt – a steel masonry fixing bolt that expands in a drilled hole as it is tightened.

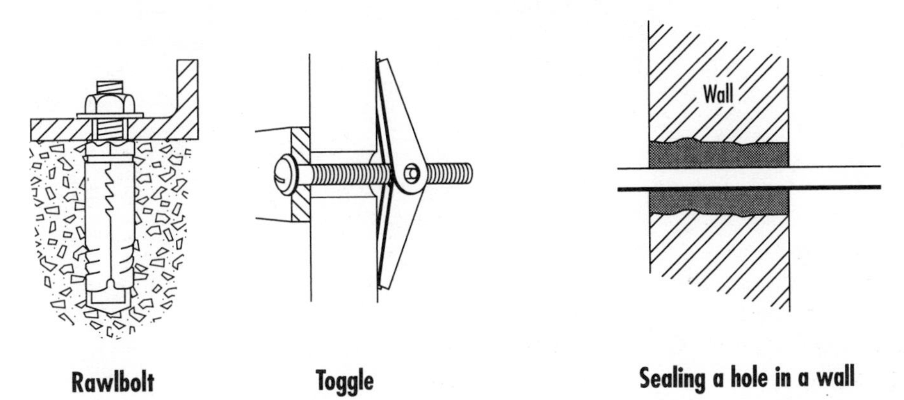

Rawlbolt **Toggle** **Sealing a hole in a wall**

Second-fix stage – a building operation stage following the first-fix stage whereby the contractor/subcontractors are able to complete their internal installations (e.g. plumbers fit boilers, radiators and sanitary appliances, carpenters fit doors, skirting and joinery fittings; electricians make connections to consumer units, light points, switches, socket outlets etc).

Site plan – a plan that is used to identify the positions of proposed buildings for setting out purposes.
Spirit level – a tool comprising several glass tubes filled with liquid and enclosing an air bubble and used for showing vertical and horizontal alignment.
Spring toggle – a plate steel spring which is pivoted on a steel nut and used for obtaining a fix in a cavity wall of light constructional strength, such as plasterboard.
Switch – a mechanical device capable of making and breaking current flowing through a circuit. A linked switch is designed to break all poles simultaneously; a time switch is one that operates through a contact or contactor; switch fuses and fused switches are ones that carry fuses and the terms switchgear and switchboard describe an assembly of switches.
Tap – a carbon steel tool having an external cut thread used for making an internal thread in a metal hole, often drilled a smaller size than the tap.
TPN – an abbreviation meaning triple-pole and neutral.
Water level – a piece of apparatus consisting of two calibrated glass tubes connected by a flexible hose which is filled with water.

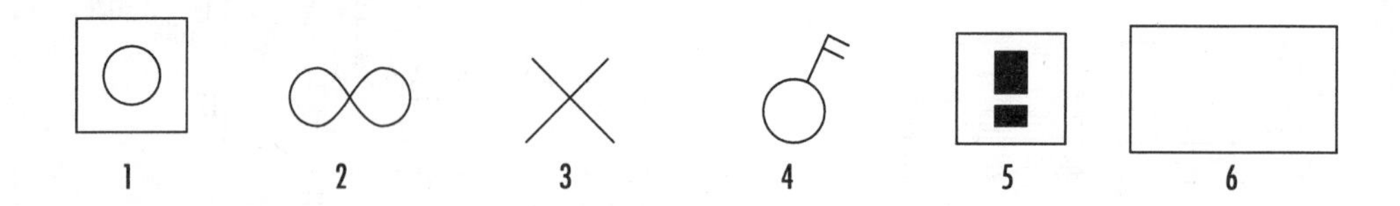

BS3939 wiring diagram symbols

1 Restricted access push button
2 Fan
3 Lighting point
4 Two-pole, 1-way switch
5 Automatic fire detector
6 Distribution board

TOPIC 5

Procedures and Practices for Work Preparation and Completion

ASSESSMENTS 5.1 – 5.3

Time allowed: 1 hour

Instructions

* You should have the following:

  Question Paper
  Answer Sheet
  HB pencil
  Metric ruler

* Enter your name and date at the top of the Answer Sheet.
* When you have decided a correct response to a question, on the Answer Sheet, draw a straight line across the appropriate letter using your HB pencil and ruler (see example below).
* If you make a mistake with your answer, change the original line into a cross and then repeat the previous instruction. There is only one answer to each question.
* Do not write on any page of the Question Paper.
* Make sure you read each question carefully and try to answer all the questions in the allotted time.

Example:

a	400 V	**a**	400 V
~~**b**~~	315 V	~~**b**~~ (crossed)	315 V
c	230 V	**c**	230 V
d	110 V	~~**d**~~	110 V

ASSESSMENT 5.1

1. What is the name of the measuring instrument shown below in Fig. 1?
   - **a** engineering compasses
   - **b** outside calliper
   - **c** vernier micrometer
   - **d** pointed scriber.

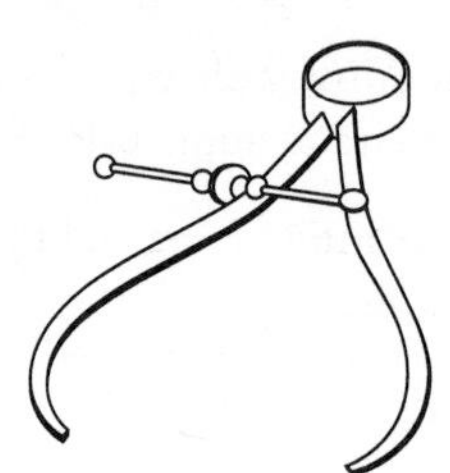

Fig. 1

2. The measuring instrument in Fig. 1 is used for finding an object's
   - **a** diameter
   - **b** circumference
   - **c** constant (π)
   - **d** radius.

3. The easiest device to use for horizontal alignment around the interior of a large empty building that has an uneven floor is called a
   - **a** water level
   - **b** spirit level
   - **c** straight edge
   - **d** chalk line.

4. Which one of the following tools is likely to be the first choice when securing a piece of work?
   - **a** mole grips
   - **b** adjustable spanner
   - **c** wrench
   - **d** pliers.

5. What is the name of the measuring instrument shown below in Fig. 2?
   - **a** portable hand vice
   - **b** cable gauge
   - **c** spark plug tester
   - **d** vernier micrometer.

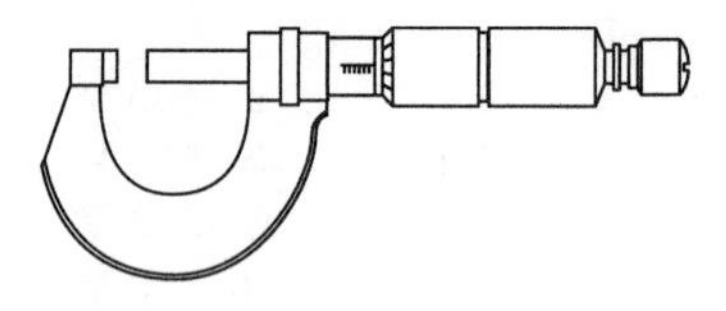

Fig. 2

6. Which one of the following suitably describes an employee's time sheet?
   - **a** a record of activities at work
   - **b** a list of daytime journeys
   - **c** a weekly total of hours worked
   - **d** a daily programme of work.

7. The term **daywork** is used to describe
   - **a** time and materials for work outside contract conditions
   - **b** work left unfinished on the day it commenced
   - **c** first stage fixing of a contractor's work
   - **d** general work to be completed during normal hours.

8. If a bulk delivery of materials arrives on site from a wholesaler, it should immediately be
   - **a** dispatched to the required area of work
   - **b** taken to the stores and protected
   - **c** checked for missing items by the driver
   - **d** checked against a copy of the original order.

9. All the following are reasonable precautions for the safe keeping of electrical materials on a construction site, *except*
   - **a** an intruder alarm
   - **b** a guard dog
   - **c** a lockable hut
   - **d** a security light.

10. Which one of the following electrical items does *not* need protection during the final connection stage of wiring?
    - **a** electric clock
    - **b** fixed appliance
    - **c** ceiling rose
    - **d** glass luminaire.

11. Which one of the following BS3939 installation location symbols in Fig. 3 is a **two-pole, one-way** switch?

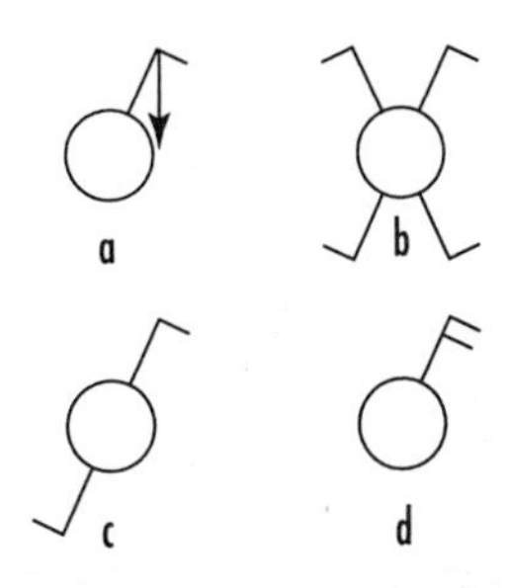

Fig. 3

12. In Fig. 4, how many live conductors pass through the point marked *XY*?
    - **a** 8
    - **b** 6
    - **c** 4
    - **d** 2.

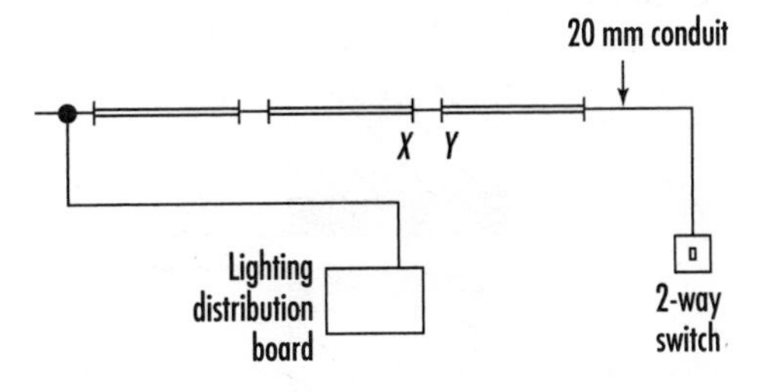

Fig. 4

13. A **switch** in which the fuse link or fuse carrier forms the moving contact when operated is called a
    - **a** time switch
    - **b** switch fuse
    - **c** fused switch
    - **d** linked switch.

14. Name the type of diagram shown in Fig. 5 below
    - **a** wiring diagram
    - **b** circuit diagram
    - **c** block diagram
    - **d** schematic diagram.

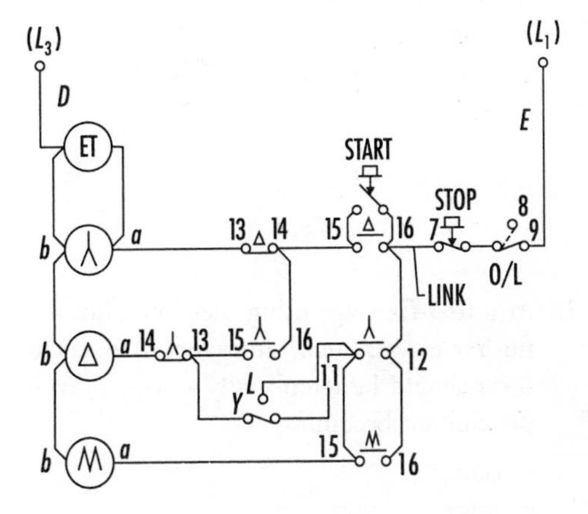

Fig. 5

15. With reference to Fig. 6, which one of the following materials is suitable for making good the large hole in the wall?

a hessian
b cement
c grout
d filler.

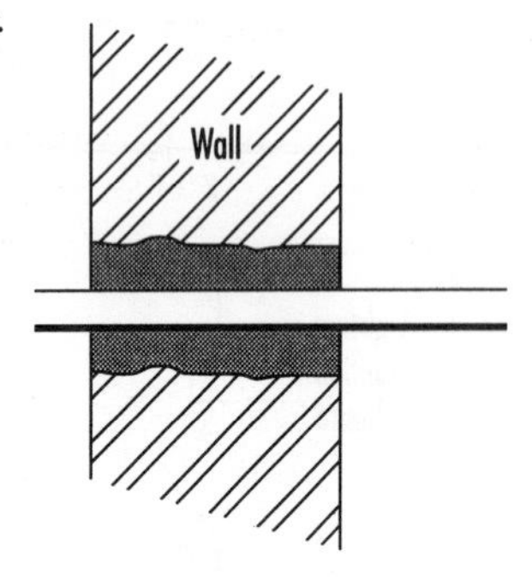

Fig. 6

16. Which one of the following is *not* classified as a marking out tool?

a scriber
b centre punch
c compasses
d micrometer.

17. Which graphical location symbol in Fig. 7 represents an automatic fire detector?

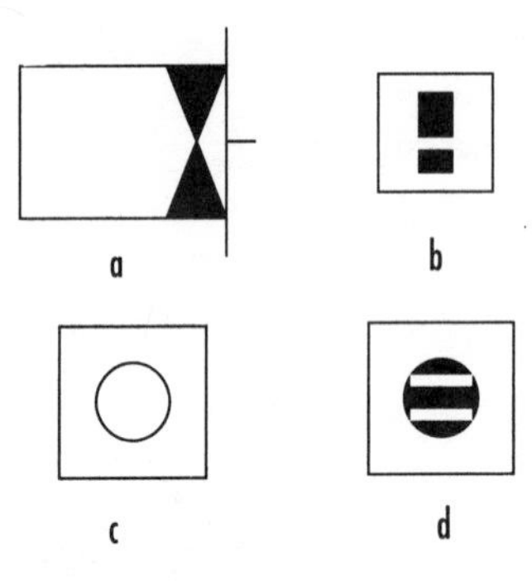

Fig. 7

18. At a **first-fix stage** using metal conduit as a flush wiring system, entry holes in enclosures should be temporarily sealed to avoid the conduit becoming

a damp
b loose
c blocked
d corroded.

19. A **spring toggle** is commonly used for fastening an object or an accessory to a

a plaster board or surface of low structural strength
b brick cavity wall that has a weak cement layer
c structure with a large hole that cannot be easily sealed
d metal frame to provide an additional earth connection.

20. All the following connections are allowable omissions from being accessible for inspection, testing and maintenance, *except* a

a compound-filled joint
b cold tail heating element
c lampholder termination
d joint formed by soldering.

21. What is the actual wall distance between socket outlet A and socket outlet B in Fig. 8?

a 9.65 m
b 9.50 m
c 9.45 m
d 9.35 m.

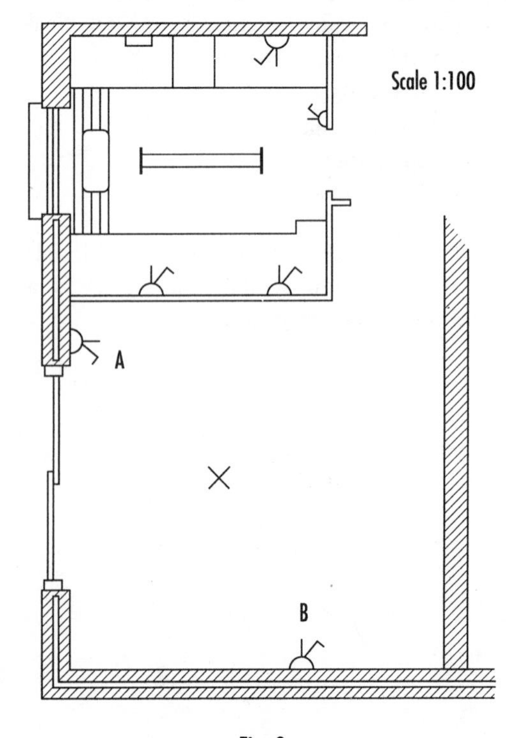

Fig. 8

22. When a black enamel steel conduit is erected outside, any bare metal should be painted to stop corrosion using

a zinc coating
b roof pitch
c clear tallow
d black emulsion.

23. The **Rawlbolt** shown in Fig. 9 obtains its fixing in the wall by means of

a contraction
b expansion
c gravity
d friction.

Fig. 9

24. A wiring system which passes through the fabric of a building should be suitably sealed to avoid the possibility of

a spread of fire
b structural damage
c rodent movement
d unwanted ventilation.

25. Which one of the following materials will *not* cause corrosion of unprotected metalwork when used in a damp situation?

a lime
b fibreglass
c plaster
d cement.

ASSESSMENT 5.1

Answers, Hints and References

1. **b** *See* Ref. 21, book 1, p 18.
2. **a** *See* Ref. 21, book 1, pp 18–19.
3. **a** *See* Ref. 21, book 1, pp 80–81.
4. **a** *See* Ref. 18, pp 79–80.
5. **d** *See* Ref. 21, book 1, p 21.
6. **c** *See* Ref. 12, p 41, Fig. 3.5.
7. **a** *See* Ref. 12, pp 39–40, Fig. 3.4.
8. **d** *See* Ref. 12, p 40.
9. **b** This is an extreme measure and is often accompanied with a patrol guard.
10. **c** It is part of the normal wiring.
11. **d** *See* Ref. 12, p 5, Fig. 1.1.
12. **c** *See* Ref. 12, p 45, Fig. 3.8.
13. **c** *See* Ref. 12, p 3.
14. **d** *See* Ref. 12, p 104, Fig. 5.3.
15. **b** *See* Ref. 4, section 527.
16. **d** *See* Ref. 21, book 1, p 20.
17. **b** *See* Ref. 12, p 5, Fig. 1.1.
18. **c** It is good practice.
19. **a** *See* Ref. 21, book 1, p 96.
20. **c** *See* Ref. 4, regulation 526–04–01.
21. **d** Remember if using cm, scale is 1 cm = 1 m.
22. **a**.
23. **b** *See* Ref. 21, book 1, pp 106–107.
24. **a** *See* Ref. 18, pp 79–80.
25. **b**.

ASSESSMENT 5.2

1. What is the name of the part marked X on the micrometer shown in Fig. 1?

   a spindle
   b barrel
   c thimble
   d anvil.

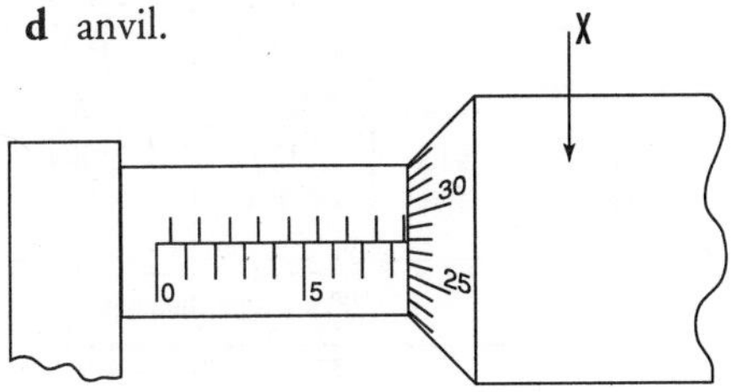

Fig. 1

2. In Fig. 1, the micrometer's reading is

   a 8.78 mm
   b 8.78 cm
   c 8.78 μm
   d 8.78 nm.

3. On a site drawing having a scale of 1:50, a wall measurement was taken with a metric ruler and found to be 20.5 cm. This means that the real length of the wall is

   a 1025 m
   b 102.5 m
   c 10.25 m
   d 1.025 m.

4. Which one of the following is used for the initial marking-off point of a long vertical conduit drop?

   a spirit level
   b water level
   c plumb line
   d straight edge.

5. Which one of the following tools is used for marking straight lines on the surface of metal?

   a 2H pencil
   b scriber
   c centre punch
   d nail set.

6. Unavoidable work outside the scope of a contract is called

   a flexiwork
   b overtime work
   c nightwork
   d daywork.

7. In Fig. 2, how many countersunk woodscrews are needed to secure the work properly?

   a 12
   b 10
   c 6
   d 4.

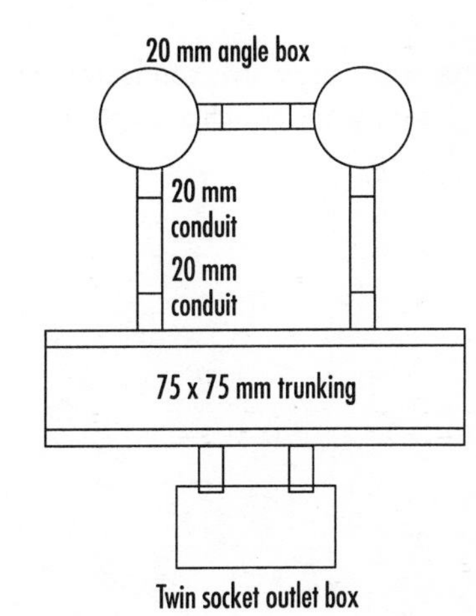

Fig. 2

8. What is the name of the BS3939 installation location symbol shown in Fig. 3?

   a kilowatt hour energy meter
   b automatic fire detector
   c manual operated call point
   d restricted access push button.

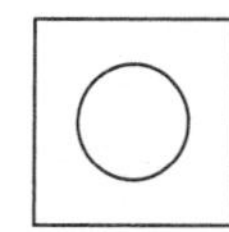

Fig. 3

9. In Fig. 4, what does the abbreviation **TPN** mean?

   a triple pole and neutral
   b three-phase and neutral
   c two-pole and neutral.
   d two-phase and neutral.

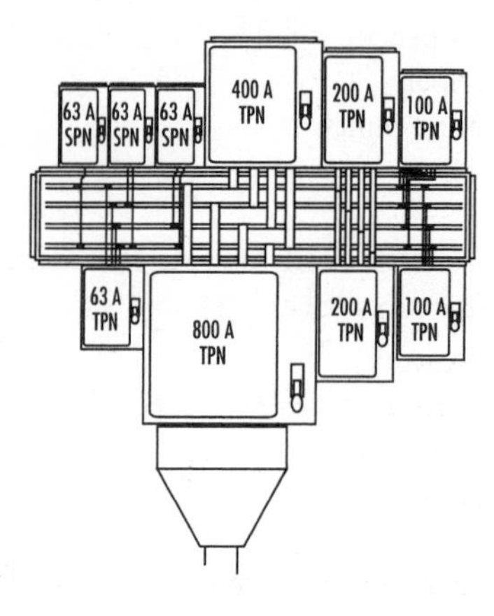

Fig. 4

10. How many live conductors are needed to supply the switchgear shown in Fig. 4?

   a 8
   b 6
   c 4
   d 3.

11. One of the reasons for writing a **job report** is to

   a see who was the supplier of material
   b provide information about how it was done
   c register any accidents to the local authority
   d keep the designer informed of progress.

12. What type of diagram is shown in Fig. 5?

   a block diagram
   b wiring diagram
   c schematic diagram
   d layout diagram.

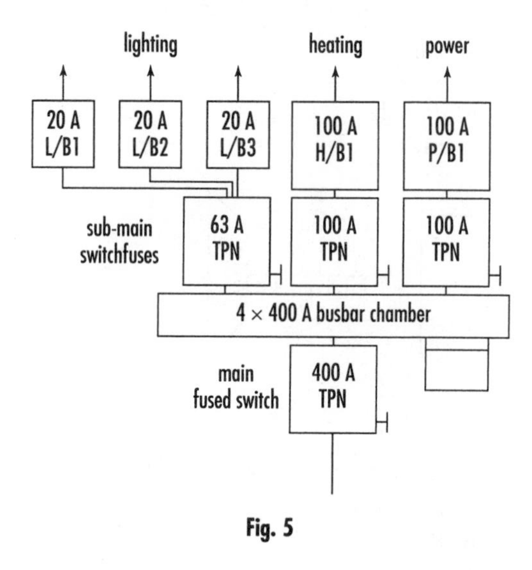

Fig. 5

13. Where a wiring system passes through the fabric of a building, any hole made is required to be suitably sealed to avoid the possibility of

   a spread of fire
   b structural distortion
   c rodent movement
   d unwanted ventilation.

14. In Fig. 6, it would be impossible to operate the load if

a only one of the start buttons is pressed
b the fuse F_2 is changed for a solid link
c L_1 and L_2 supply leads were interchanged
d the connection link marked XY is removed.

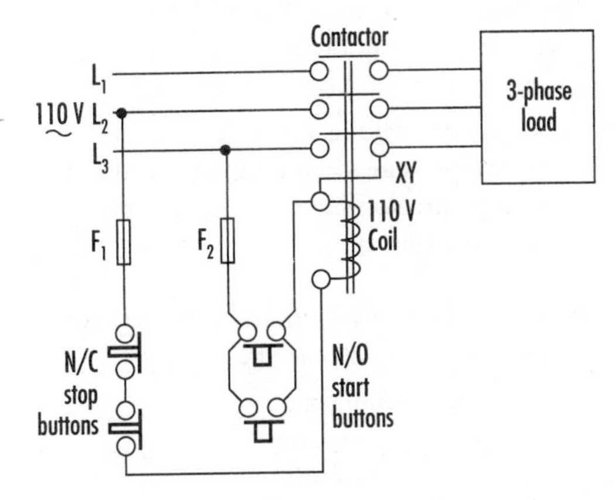

Fig. 6

15. The type of diagram in Fig. 7 showing bonding and earthing arrangements is best described as a

a layout diagram
b line diagram
c wiring diagram
d connection diagram.

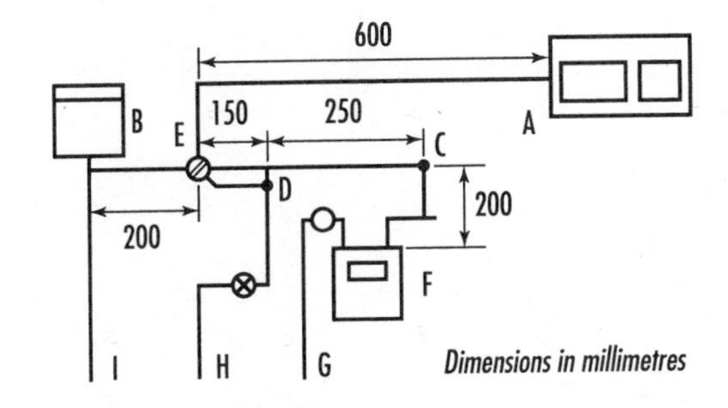

Code	Description
A	Consumer unit
B	Energy meter
C	Gas pipe connection
D	Water pipe connection
E	Earthing terminal
F	Gas meter
G	Gas intake pipe
H	Water intake pipe
I	Service cable

Fig. 7

16. In Fig. 7 earth/bonding labels are required on all of the following connection points *except* on

a A
b C
c D
d E.

17. Which of the following two materials when brought into contact with each other are likely to cause corrosion?

a rubber and zinc
b plastic and lead
c copper and porcelain
d brass and aluminium.

18. Which one of the following methods is preferred when trying to fix an accessory to a section of plaster board which has no firm backing?

a spring toggle
b Rawlplug
c loden anchor
d self-tapping screw.

19. Which one of the following woodscrews in Fig. 8 is a coach screw?

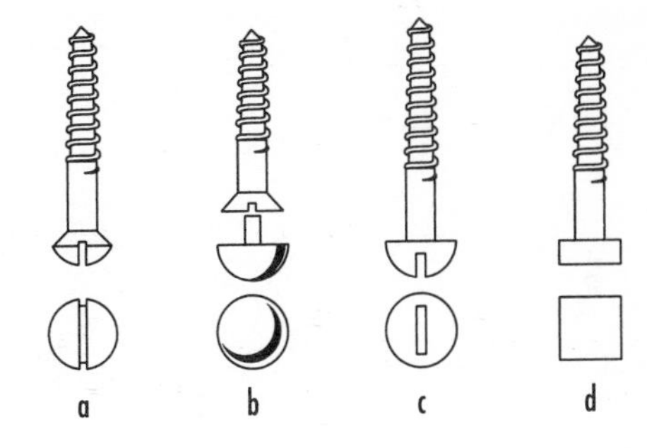

Fig. 8

20. The main reason for making a pilot hole in wood when using a screw fixing is because it will

a avoid damage to the wood and screw
b allow the screw to pierce the wood at 90°
c test the wood for its hardness property
d keep the screw flush with the surface.

21. A **bolster chisel** is an ideal tool for

- **a** stripping heavy wallpaper
- **b** cutting away metal swarf
- **c** lifting wooden floor boards
- **d** chasing stone/brick walls.

22. Fig. 9 shows the fabrication of a 90° bend in metal trunking. What is the name of the piece of metal labelled X?

- **a** side plate
- **b** blank plate
- **c** template
- **d** fishplate.

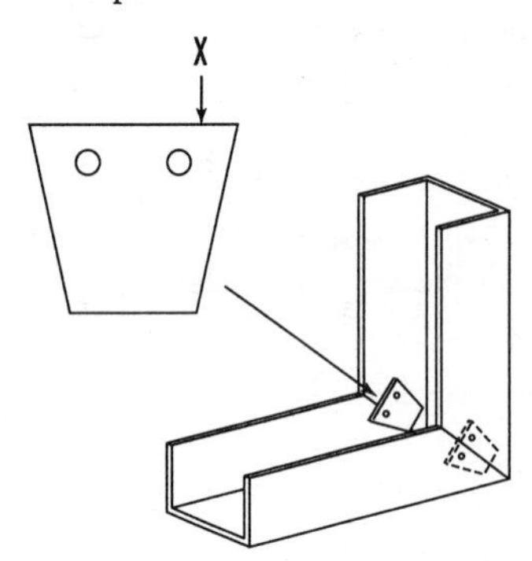

Fig. 9

23. Without additional external lugs, a PVC conduit box is allowed to support a mass of

- **a** 12 kg
- **b** 5 kg
- **c** 3 kg
- **d** 1 kg.

24. Which one of the following types of lamp needs disposing of with care in a dry container, to avoid the risk of fire?

- **a** general lighting service lamp
- **b** low pressure sodium vapour lamp
- **c** high pressure mercury vapour lamp
- **d** fluorescent lamp.

25. Before disposing of transformer oil, advice should be sought from the

- **a** product manufacturer
- **b** electrical contractor
- **c** building contractor
- **d** local fire brigade.

ASSESSMENT 5.2

Answers, Hints and References

1. **c** *See* Ref. 21, book 1, p 21.
2. **a** (8.0 + 0.5 + 0.28) *See* Ref. 21, book 1, p 21.
3. **c** 20.5/2 = 10.25.
4. **c** *See* Ref. 19, p 11.
5. **b** *See* Ref. 21, book 1, p 22.
6. **d** *See* Ref. 12, pp 39–40.
7. **b** Two for the boxes and four for the trunking.
8. **d** *See* Ref. 1, p 6.
9. **a** *See* Ref. 1, p 4.
10. **c** L_1, L_2, L_3 and N are all live conductors.
11. **b**.
12. **a**.
13. **a** *See* Ref. 4, section 527.
14. **d**.
15. **a**.
16. **a** *See* Ref. 18, p 17.
17. **d** *See* Ref. 18, pp 79–80.
18. **a** *See* Ref. 21, book 1, p 96.
19. **d** *See* Ref. 21, book 1, p 90.
20. **a**.
21. **c**.
22. **d** *See* Ref. 21, book 1, p 78.
23. **c** *See* PVC conduit catalogue.
24. **b** Sodium can catch fire if wet.
25. **a** The manufacturer will know what to do.

ASSESSMENT 5.3

1. Which one of the following is the most accurate method of finding the diameter of a small cable?

   a calliper
   b ruler
   c micrometer
   d gauge.

2. Fig. 1 shows two machine shafts about to be joined together. One method of achieving horizontal alignment is to

   a place two water levels on the flanges
   b insert callipers between the flanges
   c measure the ground height of both flanges
   d insert wooden wedges between the flanges.

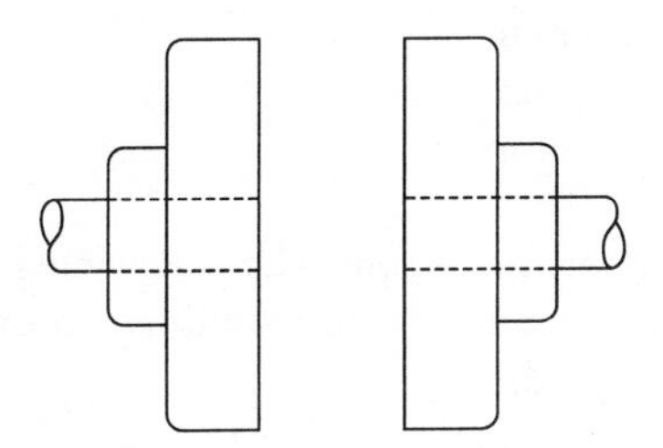

Fig. 1

3. One of the main purposes of marking out work before it is installed is that it

   a serves to reduce material wastage when the job is complete
   b provides the installer with a plan of action to complete the job
   c guarantees a perfect finish when the job is complete
   d indicates which materials are needed to complete the job.

4. An anvil, spindle, locknut and thimble are all parts of a

   a ratchet spanner
   b engineering vice
   c micrometer
   d tank cutter.

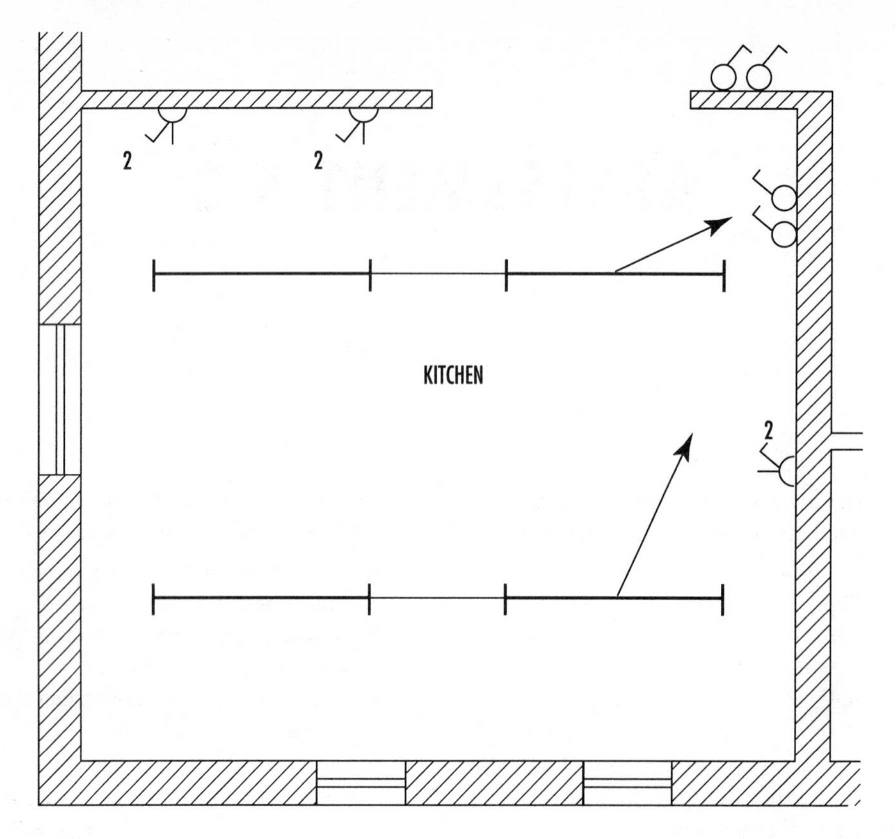

Fig. 2

5. With reference to Fig. 2 and using a scale of 1:50, what is the approximate floor area of the kitchen?

   a 32.8 m^2
   b 25.2 m^2
   c 20.9 m^2
   d 15.7 m^2.

6. In Fig. 2, what is the length of each luminaire in mm

   a 1800
   b 1500
   c 1200
   d 1000.

7. The drawing shown in Fig. 2 is called a

   a layout drawing
   b wiring drawing
   c symbol drawing
   d design drawing.

8. If the socket outlets in Fig. 2 were specified as BS1363 and a 32 A radial final circuit chosen, the minimum size of cable that could be used (ignoring correction factors) is

   a 6.0 mm^2
   b 4.0 mm^2
   c 2.5 mm^2
   d 1.5 mm^2.

9. One of the reasons why a Rawlplug fixing can fracture its surrounding masonry is because the

   a drilled hole is slightly blocked
   b type of woodscrew used is too long
   c type of woodscrew used is too large
   d wrong size screwdriver is used.

10. What is the name given to the working drawing shown in Fig. 3?

a site plan
b local plan
c block plan
d aerial plan.

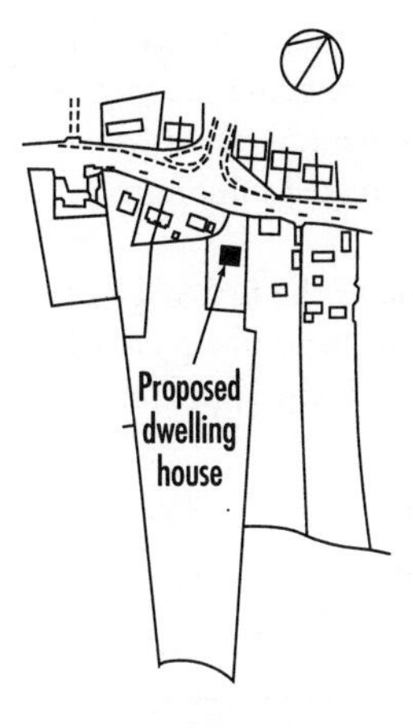

Fig. 3

11. Which one of the following BS3939 graphical location symbols shown in Fig. 4 represents a lighting distribution board?

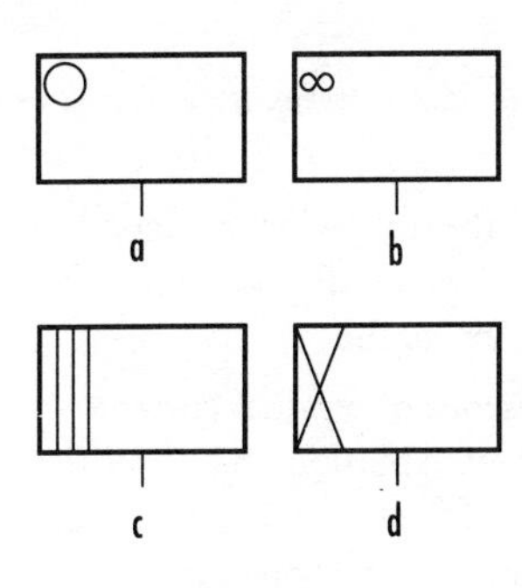

Fig. 4

12. Time and materials which are additional to a contract, causing extra work to be done are called

a daywork
b piece-work
c flexiwork
d bonus work.

13. The abbreviation **OHLS** is often associated with a type of cable and/or its accessories and means

a overhead lighting system
b zero halogen low smoke
c overall hessian layer sheath
d oil, heat and low-flame sheath.

14. If the internal cutting of a metal thread is with a **tap**, what is the component used for the cutting of an external metal thread?

a bit
b die
c screw
d taper.

15. When a new job is about to start, a material requisition list is often prepared from a

a working drawing
b pictorial drawing
c block drawing
d circuit drawing.

16. Which one of the following screws is most suitable for fixing a bathroom mirror?

a round head screw
b raised head screw
c countersunk screw
d dome head screw.

17. One method of protecting a cable concealed in a wall that runs diagonally to an outlet point, but does not meet depth and zone requirements, is to install it in

a earthed metal conduit
b rigid PVC conduit
c light gauge metal capping
d plastic mini trunking.

18. With reference to Question 17, an unprotected cable concealed in a wall and run diagonally is required to be buried to a minimum depth of

a 100 mm
b 85 mm
c 75 mm
d 50 mm.

19. Which one of the following types of fixing is *not* used for plaster board or material of low structural strength?

a spring toggle
b collapsible sleeve
c interset
d Philblock.

20. Which one of the following is *not* a consideration for using a conduit inspection accessory?

a access to modify circuit cables
b ease in which to install wiring
c access to repair circuit faults
d check on the number of circuits.

21. Which one of the following is a method of securing against unauthorized operation of a control circuit so that it cannot be switched on?

a interlock
b warning indicator
c danger notice
d fixed guard.

22. The earth electrode access point in Fig. 5 is designed to

a collect water and lower the electrode's resistance level
b allow periodic inspection and testing of the electrode
c protect the electrode against corrosion or mechanical damage
d allow air to circulate around the electrode to reduce heat.

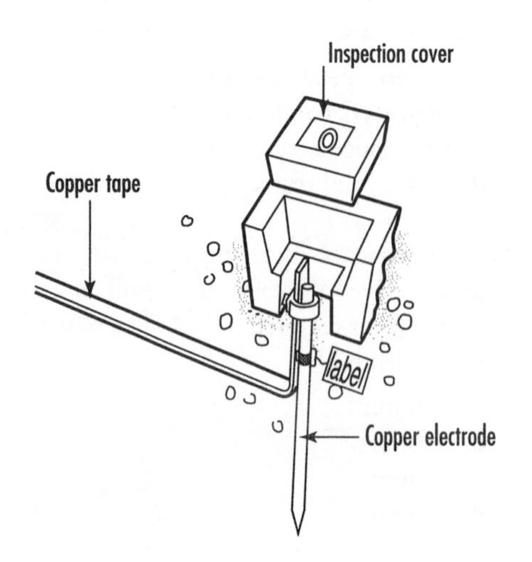

Fig. 5

23. All of the following are methods used to protect metals against corrosion *except*

a use of cathodic protection
b use of protective coatings
c use of neutral electrolyte
d use of mixed inhibitors.

24. Fig. 6 shows the first stage of making a fabricated 90° bend in a piece of 150 mm cable tray. What is the approximate length of X?

a 318 mm
b 212 mm
c 178 mm
d 96 mm.

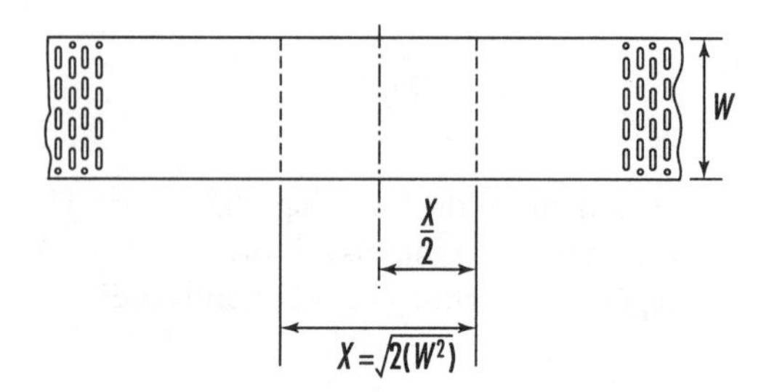

Fig. 6

25. All of the following are good reasons for using a waste disposal unit for discharge lamps, *except* the

a presence of residual current
b risk of explosion
c risk of fire
d possibility of flying glass.

ASSESSMENT 5.3

Answers, Hints and References

1. **c**.
2. **b** *See* Ref. 21, book 1, p 17.
3. **d** *See* Ref. 21, book 1, p 22.
4. **c** *See* Ref. 21, book 1, p 21.
5. **c** Note that 1 cm on your ruler is ½ m.
6. **b**.
7. **a** *See* Ref. 2, p 9.
8. **b** *See* Ref. 18, p 124, Table 8A.
9. **c**.
10. **c** *See* Ref. 2, p 9.
11. **d** *See* BS3939 *Graphical Location Symbols for Electrical Installation Drawings.*
12. **a** *See* Ref. 12, pp 39–40.
13. **b**.
14. **b**.
15. **a**.
16. **d** *See* Ref. 21, book 1, p 90.
17. **a** *See* Ref. 4, regulation 522–06–07.
18. **d** *See* Ref. 4, regulation 522–06–06.
19. **d** *See* Ref. 21, book 1, p 109.
20. **d**.
21. **a**.
22. **b**.
23. **c** *See* Ref. 3, p 63.
24. **b**.
25. **a**.

6

THE INSTALLATION OF WIRING SYSTEMS

To tackle the assessments in Topic 6 you will need to know:

- numerous definitions associated with wiring systems;
- component elements of cables with copper or aluminium conductors having different types of insulated sheath;
- working properties of different types of insulation material such as PVC, synthetic rubber, silicon rubber and magnesium oxide;
- how to choose cables for properties of particular environments such as temperature, moisture, corrosion damage by animals, sunlight, vibration and mechanical stress;
- how to determine the size of cables from the circuit design current and voltage drop, and other conditions within the installation;
- how to install and terminate PVC/PVC cables, PVC armoured cables, steel and PVC conduit, and steel and PVC trunking, as well as cable trays;
- methods of support for wiring systems and the requirements for a safe installation.

DEFINITIONS

Busbar chamber – an assembly of copper bars insulated from each other and the surrounding metal enclosure used for the purpose of making supply connections to switchgear and other electrical equipment.

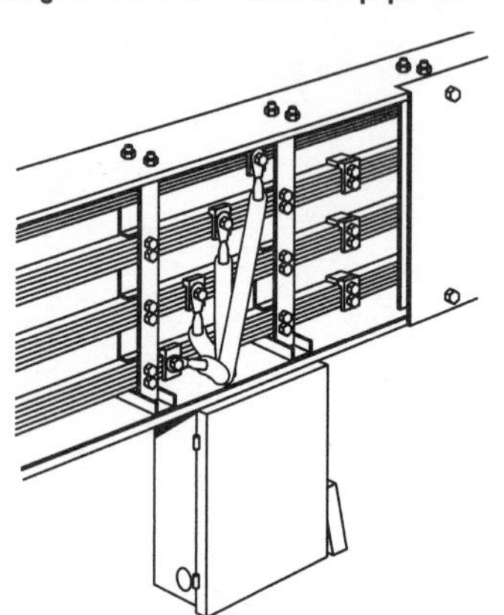

Busbar chamber

Cable – a length of insulated single-core or multicore conductor that is either solid or stranded and covered with insulation that offers additional mechanical protection.
Cable shroud – a PVC-insulated, conical-shaped glove used to cover the termination of a cable (often a mineral insulated metal sheath, MIMS, or PVC armoured cable) to give it protection against a hostile environment.

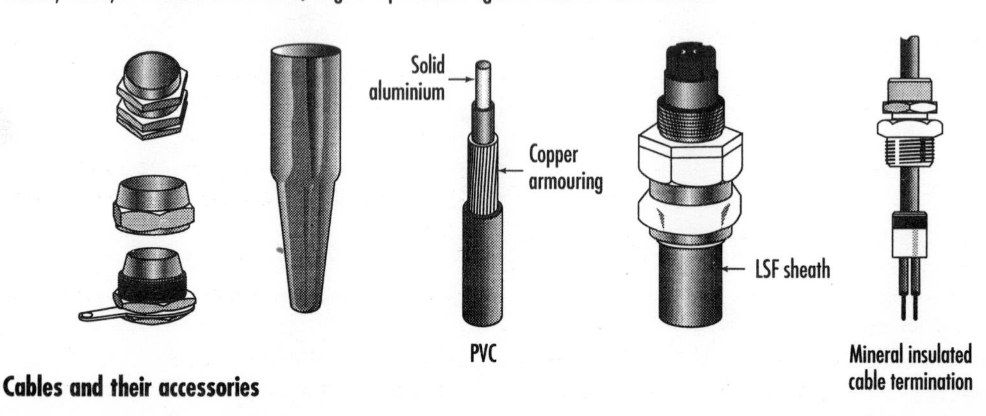

Cables and their accessories

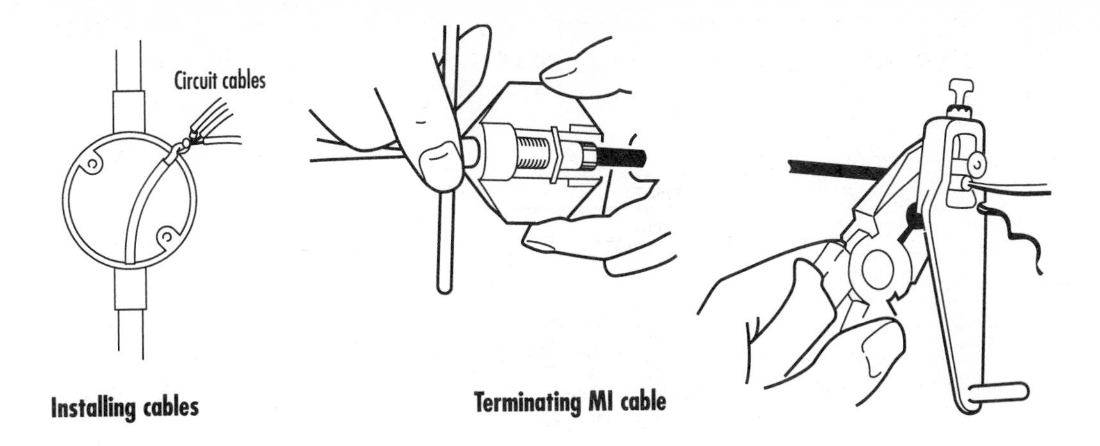

Installing cables **Terminating MI cable**

Category circuit – a circuit designated for its operation at a certain voltage or disposition. A mains supply circuit is classified as a C1 circuit, whereas a telephone circuit is classified as a C2 circuit. Fire alarm circuits and emergency lighting circuits are classified as C3 circuits.
Circuit protective conductor – a conductor that connects exposed conductive parts of equipment to the main earthing terminal.
Compartmental trunking – a trunking system designed to accommodate different category circuits by having separate compartments.
Compression ring – a copper ring fitted inside a mineral insulated (MI) cable gland to provide earth continuity between the gland and the copper sheath of the cable.
Conduit – a hollow metal or plastic pipe made in various diameters and of standard length, used for the mechanical protection of insulated cable.

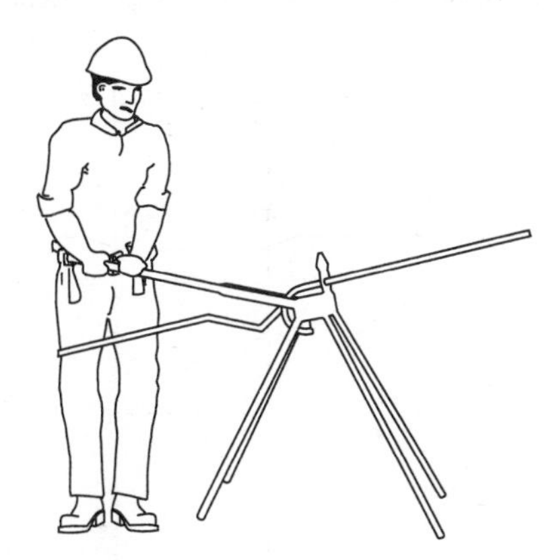

Bending conduit

Crampet – a clip used for fastening conduit to a floor surface.
Crimping tool – a tool used in the termination of an MI cable for sealing a disc to a brass screw pot.
Die – part of a tool that is used for making an external thread on the end of a conduit or piece of tubing.
Draw wire – a wire attached to circuit cables to be drawn through a conduit.
Electrolytic action – a corrosive condition set up between two dissimilar metals when they are in contact with each other.
Fire barrier – a fire resistant material such as 'Rockwool' or 'Fibreglass' often inserted between floors in the fabric of a building or even in a wiring system to stop smoke or heat penetrating other parts of the building.
Functional earthing – connection to earth necessary for the proper functioning of electrical equipment.
Grouping factor – a correction factor for cables or circuits in groups, applied where the installation conditions differ from accepted conditions.
HFOR – an abbreviation for a type of cable designed for its heat, flame and oil retarding properties.
Hygroscopic – a term frequently used to describe the condition of a substance, such as magnesium oxide, that absorbs moisture.

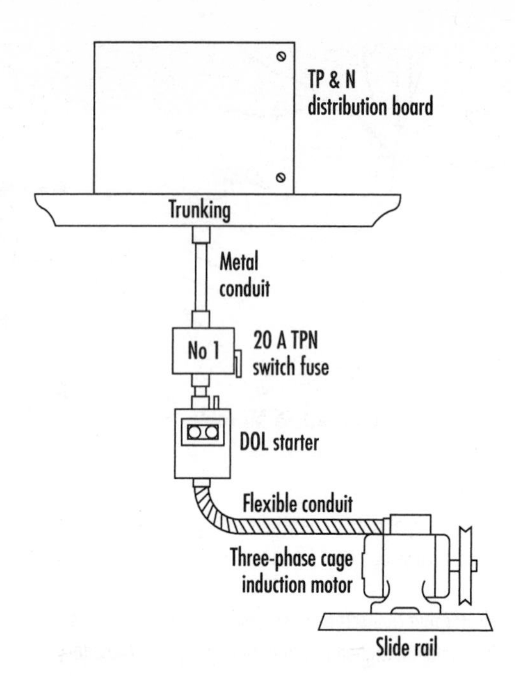

Motor installation

Interconnector – a cable or bare conductor used for the purpose of making connections inside switchgear.
Magnesium dioxide – a powdery chalk-like compound used in MIMS cable for its excellent insulation and heat properties.
OHLS – an abbreviation for cable designed to emit zero halogen and low smoke.
Pot wrench – a tool used in a MIMS termination to screw on a brass pot.
Rising main system – a wiring system that passes vertically through the fabric of a building to provide a main supply of electricity on each floor.

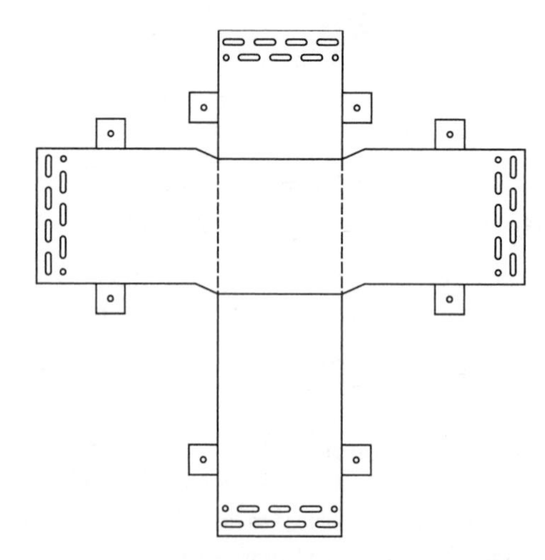

Traywork

Saddle – an accessory used for supporting and fixing metal or plastic conduit and designed for different installation conditions.
Semiconductor – a conductor material whose electrical conductivity lies somewhere between a good conductor and a good insulator.
Space factor – the ratio of the sum of the overall cross-sectional area of cables to the internal cross-sectional area of the enclosure in which they are installed.
uPVC – an abbreviation for an insulation made from polyvinyl chloride that has its plasticizing agents removed, giving it a high compression strength and high impact to resistance. Note **HRPVC** is a heat resisting PVC that can operate at 90°C.

TOPIC 6

The Installation of Wiring Systems

ASSESSMENTS 6.1 – 6.3

Time allowed: 1½ hours

Instructions

* You should have the following:

  Question Paper
  Answer Sheet
  HB pencil
  Metric ruler

* Enter your name and date at the top of the Answer Sheet.
* When you have decided a correct response to a question, on the Answer Sheet, draw a straight line across the appropriate letter using your HB pencil and ruler (see example below).
* If you make a mistake with your answer, change the original line into a cross and then repeat the previous instruction. There is only one answer to each question.
* Do not write on any page of the Question Paper.
* Make sure you read each question carefully and try to answer all the questions in the allotted time.

Example:

a 400 V	**a** 400 V
~~**b**~~ 315 V	~~**b**~~ (crossed) 315 V
c 230 V	**c** 230 V
d 110 V	~~**d**~~ 110 V

ASSESSMENT 6.1

1. The purpose of insulation around a conductor of a cable is to provide a degree of mechanical protection and also to

a improve manual handling
b prevent internal heat loss
c improve wiring flexibility
d reduce leakage current.

2. All the following are important properties of a conductor in a cable, *except*

a adequate tensile and shear strength
b installation method and support
c high resistance to corrosion
d low electrical and thermal resistance.

3. A good conductor is one that has

a high conductivity
b high resisistivity
c low conductivity
d low permeability.

4. Which one of the following terms is *not* needed to find a conductor's resistance?

a resistivity
b length
c capacitance
d cross-sectional area.

5. What is the name of the insulation material found inside the mineral insulated cable termination shown in Fig. 1?

a manganese dioxide
b potassium carbonate
c magnesium oxide
d limestone powder.

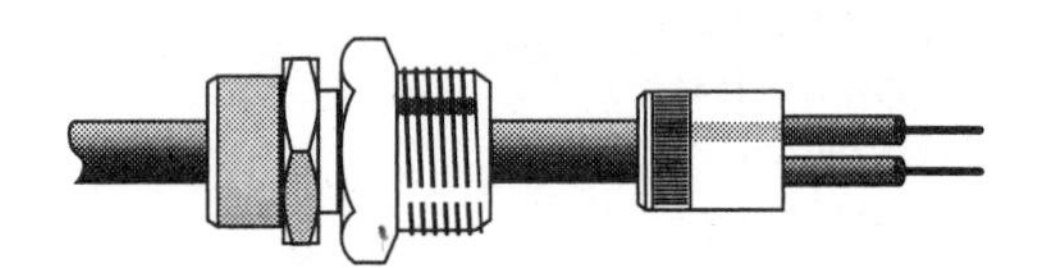

Fig. 1

6. One of the disadvantages of a mineral insulated (MI) cable is that the insulating medium may become

a hygroscopic
b hydroponic
c hydrostatic
d homogeneous.

7. Which one of the following tools would be most suitable for terminating the end of a PVC/PVC insulated cable?

a knife
b pliers
c shears
d side cutters.

8. What is the name of the cable accessory shown in Fig. 2.

a shroud
b sleeve
c glove
d sock.

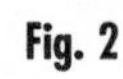

Fig. 2

9. The cable accessory shown in Fig. 2 is often used to cover a MI cable gland to protect it from the possibility of

- **a** direct contact
- **b** external influences
- **c** indirect contact
- **d** magnetic interference.

10. Which one of the following types of cable insulation is most affected by temperatures below 0°C?

- **a** butyl rubber
- **b** PVC
- **c** magnesium oxide
- **d** paper.

11. What is the minimum nominal cross-sectional area of insulated copper cables that can be used for power and lighting circuits?

- **a** 4.0 mm^2
- **b** 2.5 mm^2
- **c** 1.5 mm^2
- **d** 1.0 mm^2.

12. The factor to be applied to MI cable in order to determine the minimum internal radius of a bend is

- **a** 8
- **b** 6
- **c** 4
- **d** 2.

13. The purpose of filling an underground cable joint with epoxy resin is to prevent

- **a** entry of moisture
- **b** corrosion of the conductors
- **c** contact between conductors
- **d** mechanical movement.

14. When a cable is selected for a circuit, the first step is to find the

- **a** protective device's rating
- **b** permissible voltage drop
- **c** design current or load current
- **d** thermal constraint of the circuit protector conductor.

15. In the absence of any British Standard, the circuit voltage drop must *not* exceed

- **a** 9.6% of the nominal voltage
- **b** 6.0% of the nominal voltage
- **c** 4.0% of the nominal voltage
- **d** 2.5% of the nominal voltage.

16. The item marked X in Fig. 3 is a bonding tag that is used to give a cable termination

- **a** good mechanical strength
- **b** an external test link
- **c** a non-corrosive joint
- **d** sound earth continuity.

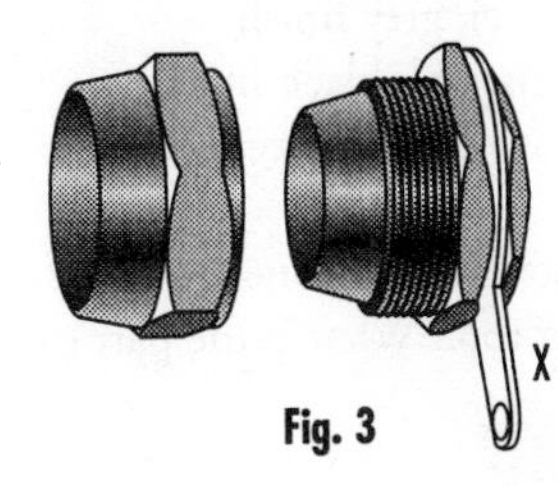

Fig. 3

17. The slotted base plate on the saddle shown in Fig. 4 is to allow the conduit to

- **a** be moved at any angle
- **b** expand and contract
- **c** move with the surface
- **d** be adjusted sideways.

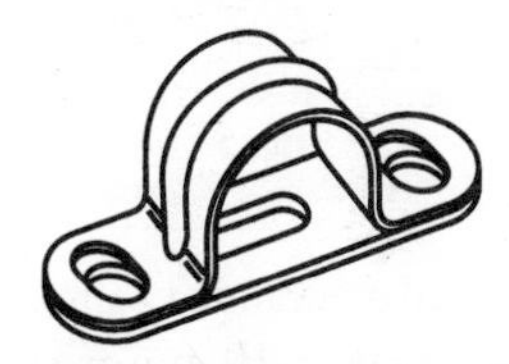

Fig. 4

18. Which one of the following wiring systems is most commonly used in domestic premises?

- **a** PVC/PVC insulated cables
- **b** silicon rubber cables
- **c** PVC steel armoured cables
- **d** mineral insulated cables.

19. When terminating a PVC armoured cable, which one of the following tools is best used to remove the stranded armouring?

- **a** wire cutters
- **b** junior hacksaw
- **c** pliers
- **d** side cutters.

20. When terminating a multicore PVC armoured cable into a wiring accessory, it is important to allow sufficient slack in the core conductors to avoid

a shrinkage and distortion
b wrongful identification
c connection errors
d mechanical stress.

21. A Class 2 metal conduit is one that has a

a hot-dipped, galvanized finish
b mild steel, grey finish
c stove enamel, black finish
d sheet steel, admiralty finish.

22. Fig. 5 shows a component part of a conduit threading tool. What is the part called?

a die
b stock
c guide
d sleeve.

Fig. 5

23. When circular stranded, single-core PVC cables of less than 10 mm overall diameter are installed in a conduit, the factor that has to be applied to establish the minimum internal radii of bends is

a 8
b 6
c 4
d 2.

24. The distance **saddle** shown in Fig. 6 is used on surfaces where there is

a an extreme cold temperature
b heavy condensation present
c constant vibration occurring
d a collection of dust and grease.

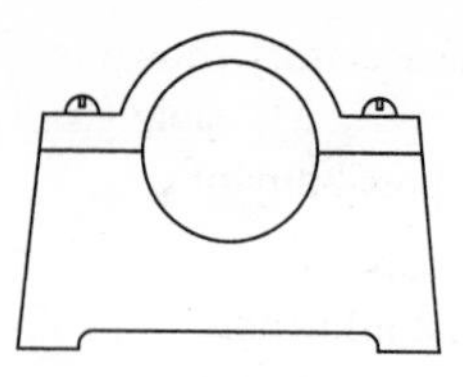

Fig. 6

25. The term deburring is a method of cleaning swarf from the ends of a conduit. This is best achieved with a

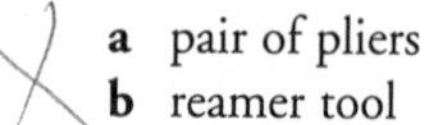

a pair of pliers
b reamer tool
c cold chisel
d large screwdriver.

26. What type of conduit connection is made at the position marked X in Fig. 7?

a running coupling
b butt joint
c locknut joint
d solid sleeve.

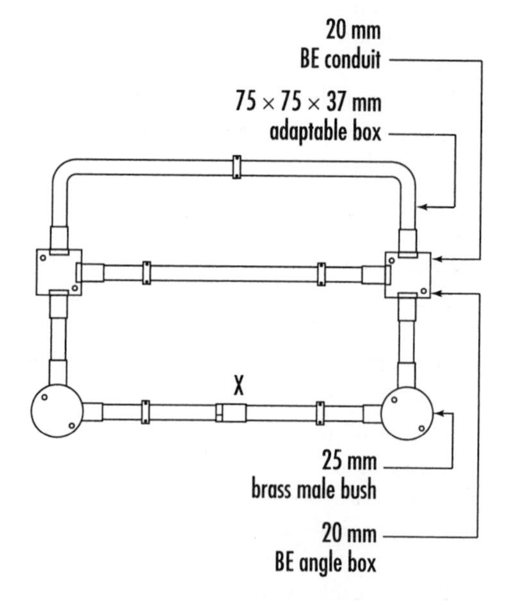

Fig. 7

27. In Fig. 7, assuming each adaptable box has the same size holes as the conduit, how many 20 mm couplings are required?

a 9
b 7
c 5
d 4.

28. Which one of the following types of wood-screw is preferred for fixing a bathroom mirror?

a raised-head screw
b round-head screw
c dome-head screw
d Pozidrive-head screw.

29. One method of protecting steel conduit against corrosion is to cover the exposed parts with

a molten tar
b zinc rich paint
c black emulsion paint
d clear varnish.

30. What is the name of the plastic conduit box shown in Fig. 8?

a four-way box
b loop-in box
c ceiling box
d back-outlet box.

31. The external lugs shown attached to the plastic box in Fig. 8 should be suitable for a maximum temperature of 60°C and also a suspended load of

a 80 kg
b 50 kg
c 20 kg
d 10 kg.

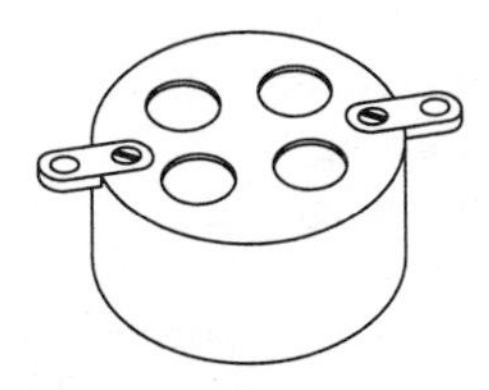

Fig. 8

32. To remove a bending spring from inside a plastic conduit you should gently pull it out and also

a tap the conduit in different places
b tap the conduit only at the empty end
c twist the spring in a clockwise direction
d twist the spring in an anticlockwise direction.

33. The ratio of the sum of the overall cross-sectional area of cables to the internal cross-sectional area of a wiring enclosure is called the

a space factor
b diversity factor
c load factor
d power factor.

34. When making a 90° bend in a conduit, the term 'bend' is equivalent to one

a double set
b bridge set
c single set
d saddle set.

35. In Fig. 9 the conduit former is designed to make a right-angle bend such that the inner radius of the bend is not less than

a 6.0 times the outside diameter of the conduit
b 4.0 times the outside diameter of the conduit
c 2.5 times the outside diameter of the conduit
d 1.5 times the outside diameter of the conduit.

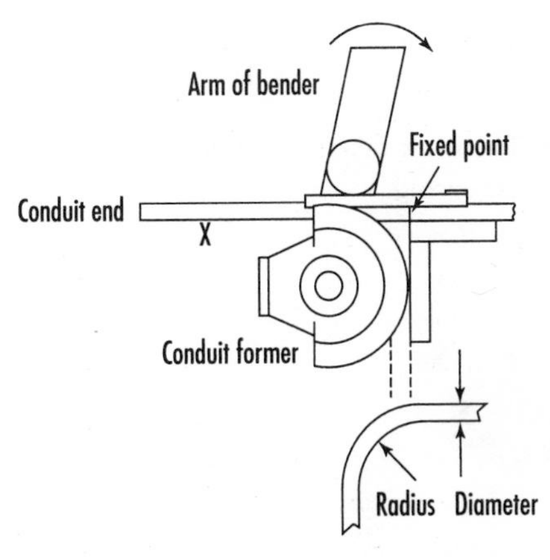

Fig. 9

36. What is the name of clip shown in Fig. 10, used to fix a conduit to a floor prior to the floor being concreted?

a crampet
b half saddle
c pipe wedge
d half-hook.

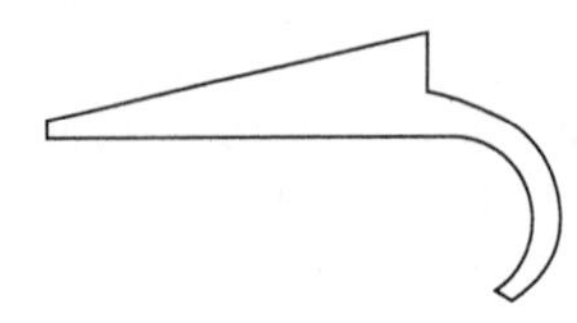

Fig. 10

37. Fig. 11 shows a copper link attached to the end section of trunking. This is to provide

a mechanical strength between lengths
b earth continuity between the lengths
c a facility to carry out earth testing
d main equipotential earth bonding.

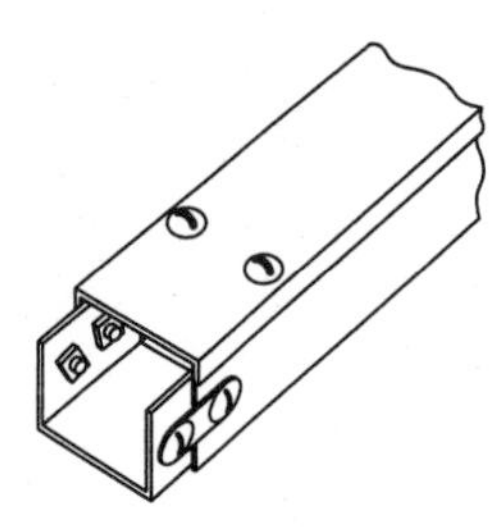

Fig. 11

38. The space factor for cables in trunking must *not* exceed

a 50%
b 45%
c 30%
d 25%.

39. One method of sealing a drilled hole in metal trunking is to fill it with an appropriate

a rubber grommet
b male bush and locknut
c conduit stop-end box
d conduit coupling and bush.

40. The use of compartmentalized trunking is to allow the accommodation of

a different category circuits
b different sizes of cable
c sheathed and unsheathed cables
d lighting and power circuits.

41. Which one of the following trunking measurements is *not* of a standard size?

a 150 mm × 75 mm
b 100 mm × 60 mm
c 75 mm × 50 mm
d 50 mm × 50 mm.

42. When wiring horizontal lengths of trunking the use of cable retainers reduces the

a stress on cables
b chance of tangled cables
c space factor for cables
d chance of trapped cables.

43. Which one of the following types of trunking is provided with its own internal circuit conductors?

a busbar trunking
b skirting trunking
c mini trunking
d dado trunking.

44. Fig. 12 shows a fire barrier placed inside a rising main busbar trunking. This is done to prevent the spread of fire to other floors and also prevent

a internal live conductors from becoming short circuited
b movement of rodents to different parts of the building
c the air temperature at the top of the system becoming too high
d objects and tools from falling into the wiring system.

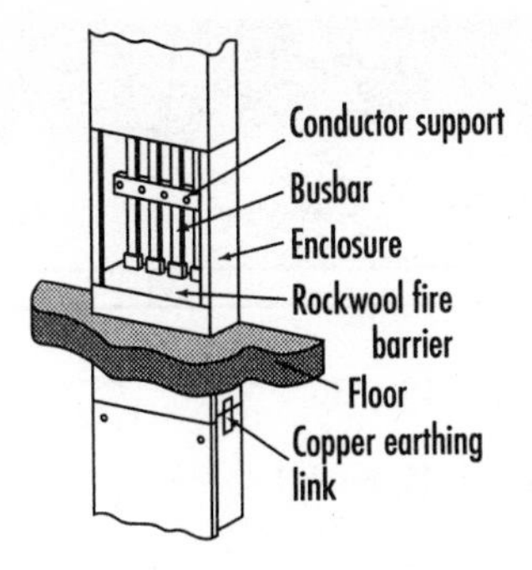

Fig. 12

45. If a lubricant is not used when cutting a thread on steel conduit it is possible to form

a a bad thread
b a crossed thread
c an oversized thread
d an oval thread.

46. When using adhesives for joining PVC conduits, care should be taken in rooms that are

a humid
b damp
c unventilated
d cold.

47. Skirting trunking has an advantage over a buried conduit system since

a no cable correction factors apply
b more socket outlets can be installed
c the building fabric is not disturbed
d it gives an exact utilization of supply.

48. The bent-over flanges on the cable tray shown in Fig. 13 provide

a additional continuity
b an attractive finish
c a means of support
d extra rigidity.

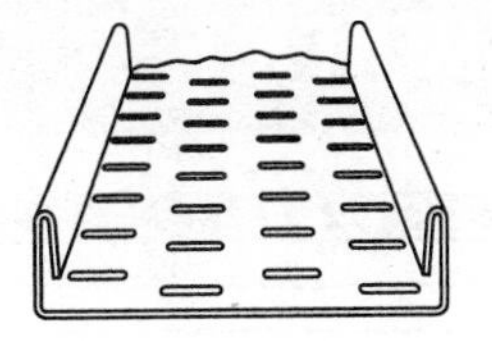

Fig. 13

49. Before PVC armoured cables are installed on a cable tray, consideration needs to be given to the

a estimated design current
b group correction factor
c type of circuit protection
d ambient temperature at 30°C.

50. In Fig. 14, the minimum depth required for drilling holes in the joists is

a 75 mm
b 50 mm
c 25 mm
d 15 mm.

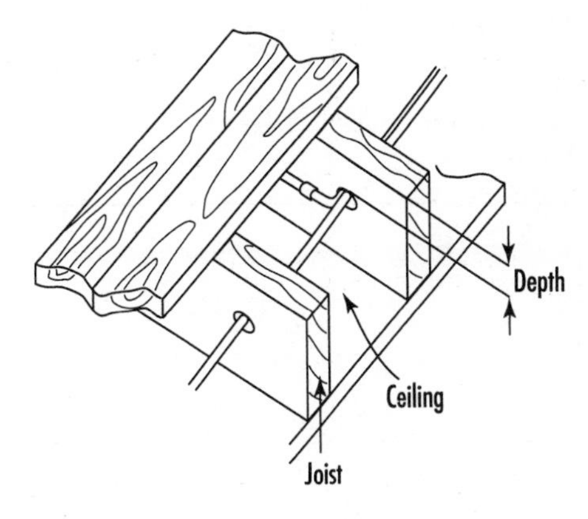

Fig. 14

ASSESSMENT 6.1

Answers, Hints and References

1. **d** *See* Ref. 2, p 32.
2. **b** It is not a related property.
3. **a** *See* Ref. 2, p 32.
4. **c** *See* Ref. 6, pp 31–32.
5. **c** *See* Ref. 1, p 69.
6. **a** *See* Ref. 1, p 69.
7. **a** *See* Ref. 19, p 14.
8. **a** *See* Ref. 2, p 36, Fig. 2.3.
9. **b** *See* Ref. 4, section 522.
10. **b** *See* Ref. 4, section 522.
11. **d** *See* Ref. 4, Table 52C.
12. **b** *See* Ref. 18, p 90, Table 4E.
13. **a**.
14. **c** *See* Ref. 4, p 175 or Ref. 1, p 73.
15. **c** *See* Ref. 4, regulation 525–01–02. NB: For a 230 V ac supply the voltage drop must not exceed $4/100 \times 230 = 9.2$ V.
16. **d** *See* Ref. 2, p 36, Fig. 2.3.
17. **d**.
18. **a**.
19. **b**.
20. **d** *See* Ref. 4, section 522.
21. **c** *See* Ref. 2, p 39.
22. **c**.
23. **d** *See* Ref. 18, p 90, Table 4E.
24. **b**.
25. **b**.
26. **a** *See* Ref. 19, p 36.
27. **b**.
28. **c** *See* Ref. 21, Book 1.
29. **b**.
30. **b** *See* manufacturers' catalogue.
31. **d** *See* manufacturers' recommendations.
32. **d** *See* manufacturers' recommendations.
33. **a** *See* Ref. 18, p 96.
34. **a**.
35. **c** *See* manufacturers' recommendations.
36. **a**.
37. **b** *See* Ref. 1, p 70, Fig. 3.47(a).
38. **b** *See* Ref. 18, p 96.
39. **a**.
40. **a** *See* Ref. 1, p 71.
41. **b** *See* Ref. 18, p 96, Table 5F.
42. **d** *See* Ref. 2, p 40, Fig. 2.7.
43. **a** *See* Ref. 2, p 41, Fig. 2.8.
44. **c** *See* Ref. 4, regulation 527–02–02.
45. **a**.
46. **c** May cause respiratory problems.
47. **c**.
48. **d** *See* Ref. 2, p 42, Fig. 2.9.
49. **b**.
50. **b** *See* Ref. 1, p 66, Fig. 3.42.

ASSESSMENT 6.2

1. Fig. 1 shows a typical BS6004 non-armoured PVC-insulated cable. Which part of the cable is designed to stop leakage currents occurring?
   - **a** circuit protective conductor
   - **b** outer sheath insulation
   - **c** conductor core insulation
   - **d** air gap between cable cores.

Fig. 1

2. In Fig. 1, which part of the cable is used to determine the cross-sectional area?
   - **a** live conductors
   - **b** protective conductor
   - **c** outer PVC sheath
   - **d** inner core PVC sheath.

3. With reference to Fig. 1, what is the name of the bare conductor marked X?
   - **a** supplementary conductor
   - **b** circuit protective conductor
   - **c** earthing conductor
   - **d** earth continuity conductor.

4. If the insulation resistance measured between the cores of a two-core cable 100 m long is 1000 MΩ, what is the insulation resistance of 50 m of the same cable?
   - **a** 4000 MΩ
   - **b** 3000 MΩ
   - **c** 2000 MΩ
   - **d** 1000 MΩ.

5. The two most common cable conductor materials used today are copper and
   - **a** brass
   - **b** nickel
   - **c** aluminium
   - **d** lead.

6. The maximum operating temperature for copper conductors with thermosetting insulation is
   - **a** 90°C
   - **b** 85°C
   - **c** 70°C
   - **d** 60°C.

7. The compression ring shown in Fig. 2 is used to provide
   - **a** conductor alignment when the gland is being tightened
   - **b** earth continuity between the gland and cable sheath
   - **c** a weatherproof seal to stop corrosion occurring
   - **d** a distance link between the back nut and gland body.

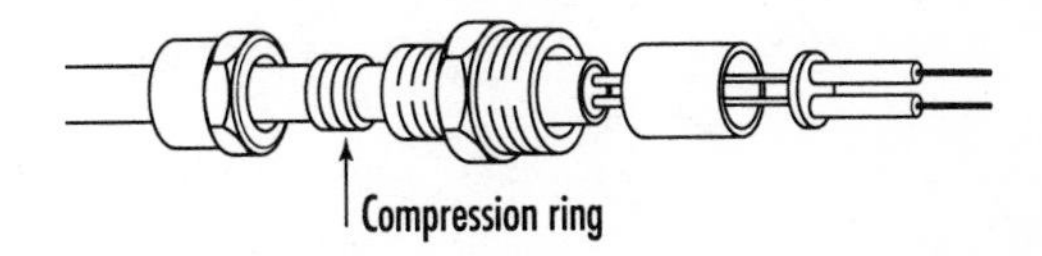

Fig. 2

8. One of the reasons for choosing an MI cable for a wiring system is that it can

a operate at a high temperature
b bend without fracturing
c withstand chemical attacks
d act as a Category 2 circuit.

9. **Magnesium dioxide** is used as the insulation material for a

a BS5467 XLPE cable
b BS6007 rubber cable
c BS5467 armoured cable
d BS6207 MI cable.

10. The shroud shown in Fig. 3 is often used to cover cable glands to protect them from

a direct contact
b indirect contact
c external influences
d magnetic interference.

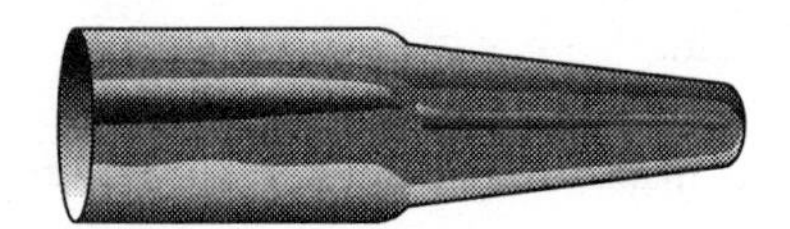

Fig. 3

11. When fixed wiring cables are bent, the bends are limited to numerical factors applied to their

a bending equipment
b cross-sectional areas
c overall diameters
d internal core radii.

12. What is the name of the tool shown in Fig. 4?

a sheath-removing tool
b pyro-cutting tool
c crimping tool
d rotary-stripping tool.

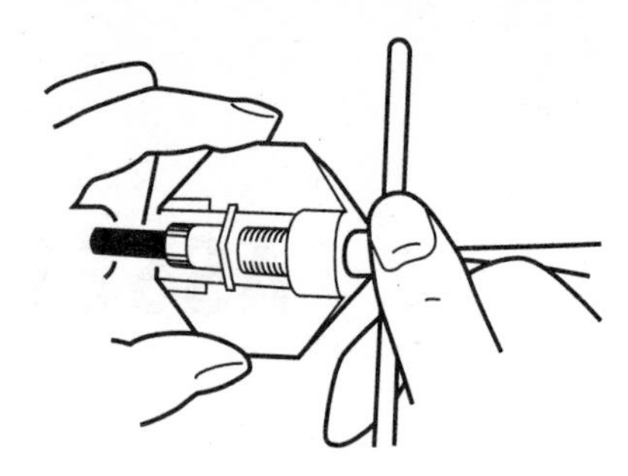

Fig. 4

13. When a cable is selected for a circuit, the first step is to find the design current and the second step is to

a determine all the relevant correction factors
b select a circuit protective device
c calculate the circuit voltage drop
d find the size of a suitable circuit conductor.

14. In Fig. 5, the wire attached to the circuit cables is called a

a fish wire
b draw wire
c pull wire
d tie wire.

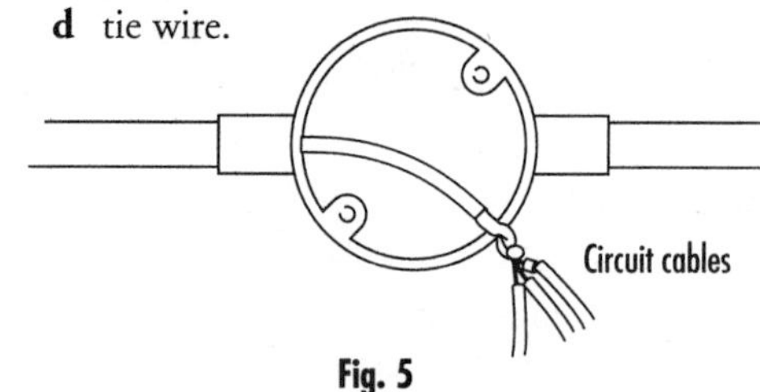

Fig. 5

15. If the circuit voltage drop is limited to 4.0%, what is the allowance for a nominal supply at 230 V?

a 9.2 V
b 6.0 V
c 4.8 V
d 2.5 V.

16. In the process of erecting steel conduit, it is necessary to make all connections tight in order to

a stop the work from moving
b prevent the ingress of dust
c protect the cables against damage
d maintain electrical continuity.

17. What is the maximum length of span between buildings for an unjointed 25 mm diameter steel conduit installed above ground?

a 3.0 m
b 2.5 m
c 2.0 m
d 1.5 m.

18. In Fig. 6, holding pliers at right angles to the copper sheath,

a reduces loss of the internal insulation
b helps to steady the termination process
c allows the sheath to be cleanly removed
d marks the position of the screw-on pot.

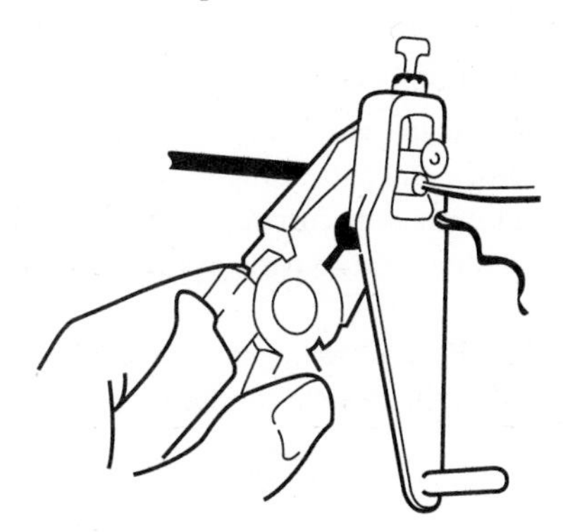

Fig. 6

19. The type of conduit recommended to be used outdoors for surface wiring is

a pliable PVC conduit
b rigid PVC conduit
c black enamel metal conduit
d galvanized metal conduit.

20. To prevent the abrasion of cables, conduit ends should be

a fitted with a coupling and locknut
b tightly screwed into accessories
c reamed with a proper deburring tool
d cut off and filed square.

21. The type of conduit set being made in Fig. 7 is called a

a bridge set
b double set
c saddle set
d single set.

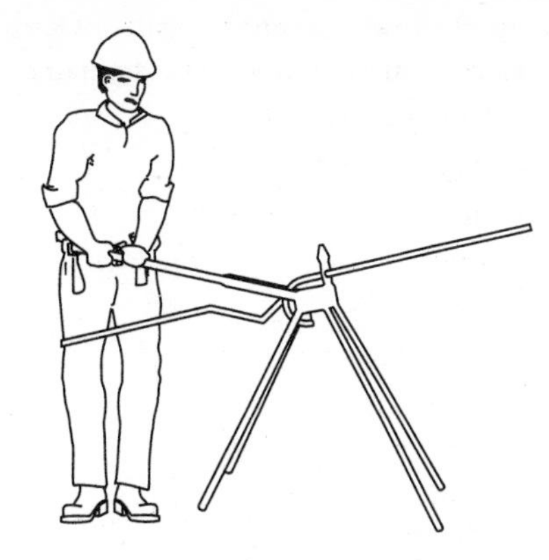

Fig. 7

22. The type of cable termination shown in Fig. 8 is called a

a soldered lug
b bolt connector
c pillar connector
d male sleeve.

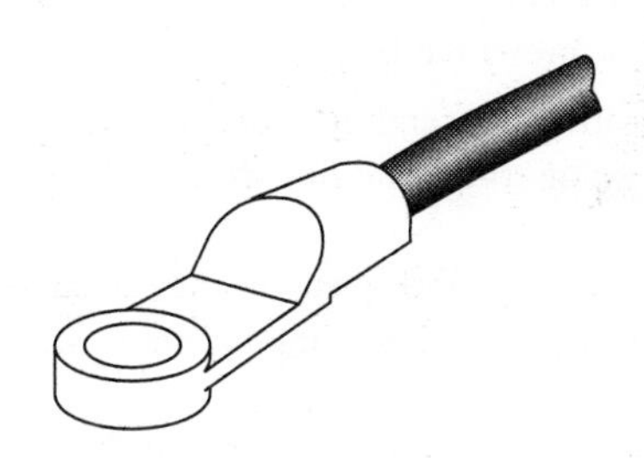

Fig. 8

23. Inspection type conduit accessories are commonly used to provide a

a degree of mechanical strength
b means for drawing-in cables
c test point for circuit cables
d link with other wiring systems.

24. When rigid metal conduit is connected to flexible metal conduit the latter should

a be constructed of waterproof material
b not be able to kink or unwind
c contain a separate protective conductor
d comprise an insulated PVC outer sheath.

25. Fig. 9 shows component parts of a conduit bending machine. What is the name of the part marked X?

a shoe
b stop
c guide
d sleeve.

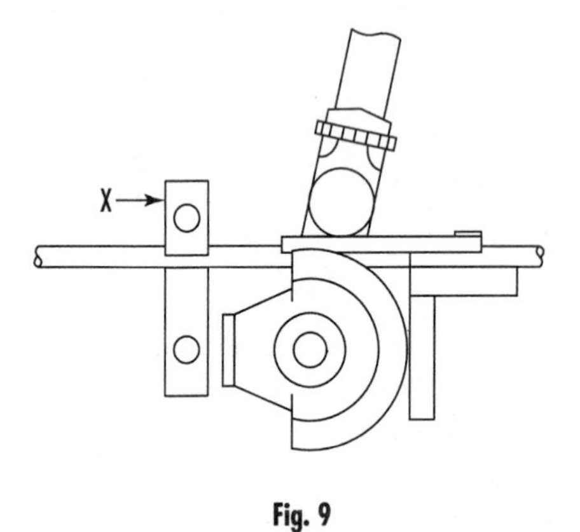

Fig. 9

26. When PVC conduit is being bent, kinking may be prevented by

a heating the conduit after it is bent
b heating the conduit while making the bend
c inserting a steel spring inside the conduit
d using a special bending former.

27. Steel conduit that is installed outdoors should preferably be

a gas tight
b painted black
c galvanized
d nickel-plated.

28. Which one of the following methods is used to allow plastic conduit to expand?

a saddles left slightly loose
b slip-joint box
c slip-joint coupling
d saddles spaced every 200 mm.

29. In Fig. 10, flexible conduit is used as a final connection because the motor

a conducts considerable heat energy
b may need to be moved or adjusted
c can be remote from its starter
d can be easily disconnected.

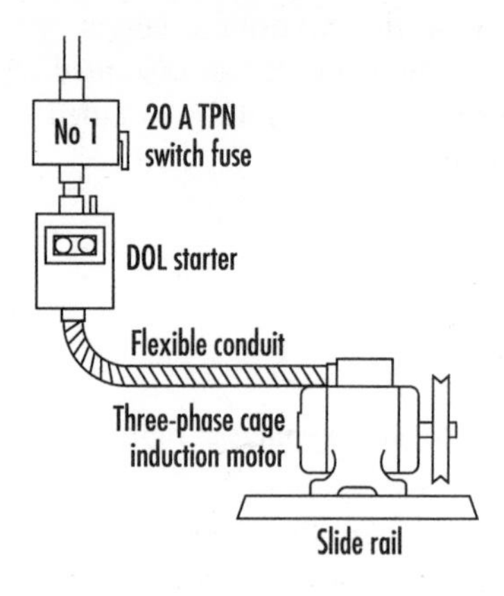

Fig. 10

30. Another reason for using flexible conduit in Fig. 10 is because the motor is not likely to

a loosen any of its foundation bolts
b loosen any part of the fixed wiring
c transmit eddy currents to the fixed wiring
d conduct heat to parts of the fixed wiring.

31. In Fig. 10, how many cables would be contained in the flexible conduit?

a 5
b 4
c 3
d 2.

32. A fire alarm and emergency lighting system are classified as

a Category 0 circuits
b Category 1 circuits
c Category 2 circuits
d Category 3 circuits.

33. The overall cross-sectional area of a 10 mm^2 PVC single core cable is 36 mm^2. What is the approximate space factor when 9 cables are placed inside 50 mm × 38 mm trunking with a trunking term of 767?

a 57%
b 50%
c 45%
d 42%.

34. A 30 A ring final circuit with no spurs is wired in single-core PVC cables and installed in PVC conduit. The number of live conductors leaving the distribution board to feed socket outlets will be

a 6
b 5
c 4
d 3.

35. Which one of the following types of cable insulation is most likely to be chosen for use in direct sunlight?

a silicone rubber
b polyvinyl chloride
c vulcanized rubber
d impregnated paper.

36. An LSF sheathed cable is one that is designed

a not to collect dust
b not to ignite and burn
c for long life expectancy
d to withstand chemical attack.

37. Flat PVC/PVC cables concealed in walls at a depth of less than 50 mm that run *diagonally* to accessory points have to

a incorporate an earthed protective conductor
b be protected by an earthed metal enclosure
c be enclosed in a protective PVC channel
d be installed only from ground level.

38. Which one of the following should not be used to fill an open entry hole left on top of an insulated consumer unit?

a coupling and bush
b closed rubber grommet
c fabricated blank plate
d plain plug.

39. What is the name of the set shown in Fig. 11, given to the horizontal tray section?

a bridge set
b saddle set
c double set
d raised set.

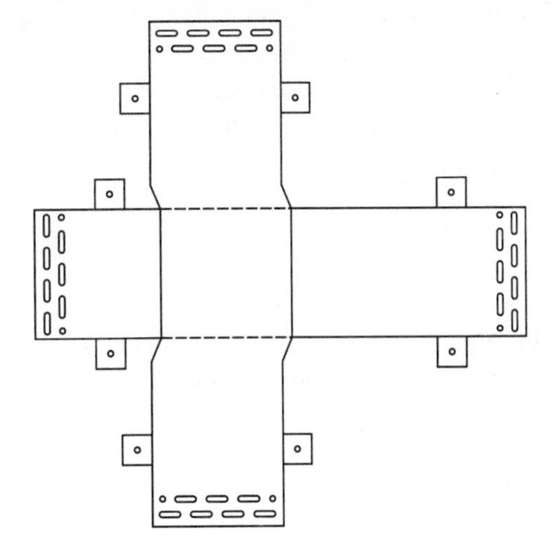

Fig. 11

40. When single-core ac cables of a particular final circuit pass from a steel trunking into a metal accessory box, they must be

a of a voltage rating at least 300/500V
b all taken through the same entry hole
c all belong to the same category circuit
d additionally sleeved to avoid damage.

41. The purpose of the flexible metal straps connecting long sections of busbar in a rising main trunking system is for

a different connections
b wiring adjustment
c mechanical movement
d thermal expansion.

42. All the following are precautions to prevent damage to cables when installing metal trunking, *except*

a use of proprietary bushes
b avoidance of sharp edges
c effective joining of sections
d connection of earth links.

43. Cornice trunking is a method of surface wiring to

a ceilings
b work benches
c skirting-boards
d architraves.

44. If a 75 mm × 25 mm trunking has a factor of 738 and a stranded 2.5 mm^2 copper cable has a factor of 11.4, approximately how many cables can be installed in the trunking?

a 65
b 50
c 43
d 29.

45. What is the conductor diameter of a solid 2.5 mm^2 single-core PVC cable?

a 3.183 mm
b 1.784 mm
c 0.795 mm
d 0.562 mm.

46. In Question 45 above, if the overall diameter of the cable is 3.9 mm, what is the thickness of the insulation, assuming it has only one insulation?

a 4.34 mm
b 3.08 mm
c 1.06 mm
d 0.22 mm.

47. One of the advantages of installing a power trunking system in a factory (*see* Fig. 12) is that it may be used

a directly from the main intake position
b without interfering with work on the floor below
c as a single-phase supply or three-phase supply
d in advance of floor supplies to machines.

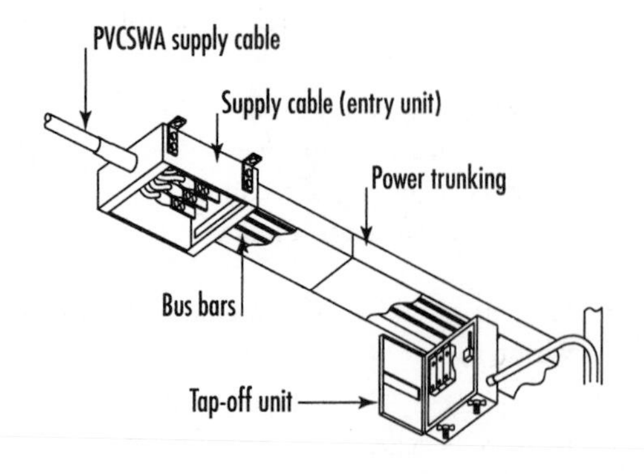

Fig. 12

48. One of the factors to be considered when cables touch each other on a cable tray is the

a current rating factor
b space factor
c ambient temperature factor
d diversity factor.

49. One of the advantages of using cable tray over trunking is its

a capacity to carry a large number of armoured cables
b wider use over different parts of a building
c ability to be bent into many different shapes
d use as a system in basements.

50. When a bend is being made in a metal tray, the formula shown in Fig. 13 is sometimes used. What is the length of the measurement shown if $r = 0.15$ m?

a 4.50 m
b 3.20 m
c 1.90 m
d 0.24 m.

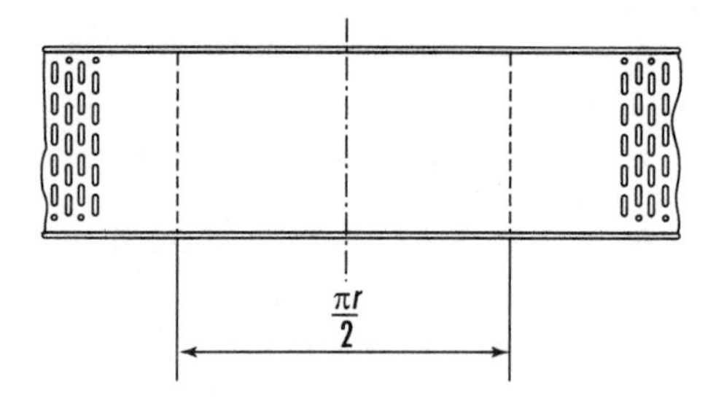

Fig. 13

ASSESSMENT 6.2

Answers, Hints and References

1. **c** *See* Ref. 2, p 32.
2. **a** *See* cable manufacturers' catalogues.
3. **b** *See* Ref. 2, p 32.
4. **c** Insulation resistance of a cable is inversely proportional to its cross-sectional area.
5. **c** *See* Ref. 2, p 32.
6. **a** *See* Ref. 2, p 36.
7. **b**.
8. **a** It is the most important factor.
9. **d** *See* Ref. 1, p 69.
10. **c** *See* Ref. 2, p 36, Fig. 2.3.
11. **c** *See* Ref. 18, Table 4E.
12. **c** *See* Ref. 19, p 17, Fig. 1.18.
13. **b** *See* Ref. 1, p 72.
14. **b**.
15. **a** 0.04×230 V = 9.2 V.
16. **d**.
17. **a** *See* Ref. 18, Table 4B.
18. **c** *See* Ref. 19, pp 17–18.
19. **d**.
20. **c** *See* Ref. 19, p 35.
21. **c** *See* Ref. 19, Fig. 2.5, p 35
22. **a**.
23. **b**.
24. **c** *See* Ref. 4, regulation 543–02–01.
25. **b**.
26. **c** *See* PVC manufacturers' instructions.
27. **c** It does not rust.
28. **c**.
29. **b**.
30. **b**.
31. **b** Three lives and PE conductor.
32. **d** *See* Ref. 4, part 2, definitions.
33. **d** $324/767 = 0.42$ approx.
34. **c**.
35. **a** It has the highest temperature rating.
36. **b** The abbreviation means low smoke fume.
37. **b** *See* Ref. 18, pp 37–38.
38. **a**.
39. **a**.
40. **b** It is to stop the possibility of electromagnetic induction.
41. **d** Busbars expand considerably when delivering full current.
42. **d** The question refers to installing the wiring system.
43. **a** *See* manufacturers' catalogues.
44. **a** *See* Ref. 18, Tables 5E and 5F.
45. **b** Use formula $d = \sqrt{(4A/\pi)}$.
46. **c** $(3.9 - 1.784)/2 = 1.058$ mm.
47. **d** *See* Ref. 2, p 41.
48. **a** *See* Ref. 4, Table 4BI.
49. **a**.
50. **d**.

ASSESSMENT 6.3

1. Fig. 1 shows a BS6004 flat twin PVC sheathed cable having circuit protective conductor. If the protective conductor is smaller than the live conductors its resistance will be

   a half of each live conductor
   b equal to each live conductor
   c equal to the sum of both live conductors
   d greater than each live conductor.

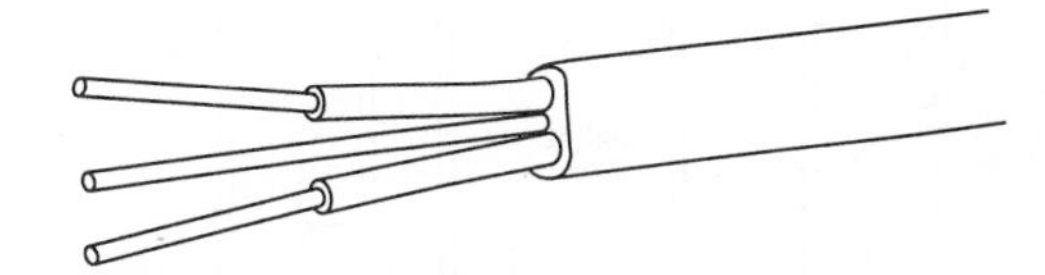

Fig. 1

2. In Fig. 1.0, if the size of the cable was 2.5 mm^2, the circuit protective conductor size would normally be

   a 2 mm^2
   b 1.5 mm^2
   c 1.0 mm^2
   d 0.5 mm^2.

3. Which one of the following is a property of a good insulator?

   a high resistance
   b high conductance
   c low resistivity
   d low impedance.

4. A cable abbreviated **uPVC** is one that has

   a no plasticized agents added
   b no property of self-recovery
   c low compression strength
   d low impact resistance.

5. Which one of the following can be classified as a semiconductor?

   a brass
   b lead
   c aluminium
   d carbon.

6. All of the following are abbreviations for types of cable sheath, *except*

   a PVC
   b LSF
   c MCCB
   d EPR.

7. The maximum operating temperature for copper cables covered with general PVC insulation is

   a 90°C
   b 85°C
   c 70°C
   d 60°C.

8. The substance **magnesium dioxide** is used in a cable as

   a an insulant
   b an electrolyte
   c a coolant
   d a conductor.

9. A 'FIRETUF' **OHLS** cable is one that is rated for its

a high capacitance value
b non-kinking qualities
c zero halogen emission
d non-electrostatic screening.

10. Which one of the following is a typical voltage designation for a common cable?

a 750/800 V
b 500/600 V
c 300/500 V
d 250/400 V.

11. In Question 10 above the first voltage indicates the voltage

a between any two conductors
b between conductors and earth
c after a voltage drop is applied
d before a voltage drop is applied.

12. What is the name of the MI termination tool shown in Fig. 2?

a pot-wrench tool
b pyro-cutting tool
c crimping tool
d sheath-removing tool.

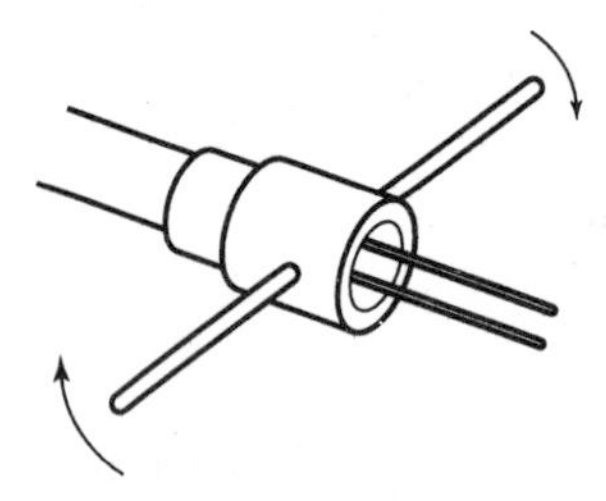

Fig. 2

13. All of the following are steps to be taken when selecting a cable for a circuit, *except* finding the

a disconnection time
b space factor
c voltage drop
d design current.

14. Which one of the following procedures is correct in order to properly select a cable for a circuit?

a $I_n \leqslant I_b \leqslant I_c$
b $I_g \leqslant I_b \leqslant I_t$
c $I_z \leqslant I_c \leqslant I_b$
d $I_b \leqslant I_n \leqslant I_z$.

15. Fig. 3 shows trunking and conduit for a motor installation. If the total cable term for the cables run in the trunking is calculated to be 3150, which one of the following trunking sizes is the smallest that can be used to keep within a space factor of 45%?

a 225 mm × 38 mm trunking size; 3474 trunking term
b 150 mm × 50 mm trunking size; 3091 trunking term
c 100 mm × 75 mm trunking size; 3189 trunking term
d 150 mm × 38 mm trunking size; 2999 trunking term

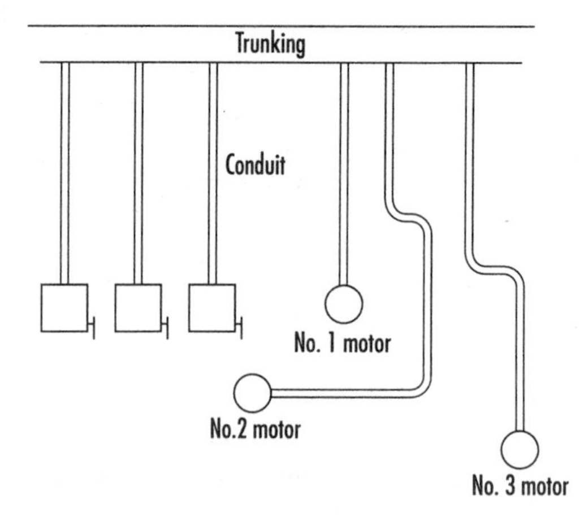

Fig. 3

16. In Fig. 3 the number of right-angle bends in the conduit running to motor no.2 would be counted as

a 3
b 2
c 1.5
d 1.

17. MIMS cable is sometimes provided with an overall PVC sheath to

a make it bend more easily
b reduce its heat conduction
c protect it from sunlight
d prevent it from corroding.

18. The maximum distance between saddles for 25 mm plastic conduit running vertically is

a 2.0 m
b 1.2 m
c 0.6 m
d 0.3 m.

19. A stop and former are two essential components of a

a conduit-bending machine
b tray-bending machine
c fixed-pillar drill
d cable-crimping tool.

20. Why is it necessary to apply a factor to the overall diameter of a cable to determine its minimum bend radius?

a it allows the wiring to look tidy
b it improves the space factor
c it avoids internal core stress
d it avoids kinking of the sheath.

21. In Fig. 4, the recommended maximum distance to install either a conduit solid elbow or solid tee from an outlet box is

a 2.0 m
b 1.5 m
c 1.0 m
d 0.5 m.

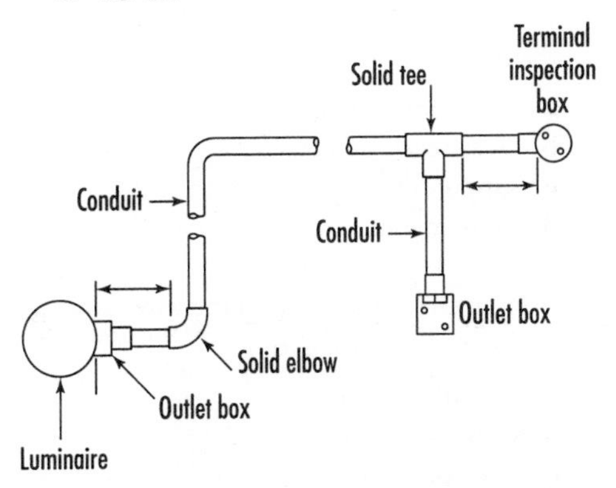

Fig. 4

22. Eight single PVC cables are installed in a conduit. If each cable had a factor of 27 and the conduit had a factor 460, how many more similar cables can the conduit safely accommodate?

a 12
b 9
c 6
d 3.

23. When terminating a single-phase conductor of a yellow phase supply into a single-pole switch it should be

a left the same colour
b labelled with the letter Y
c labelled with code L2
d identified by red sleeving.

24. In damp situations, aluminium core cables should not be terminated into brass connections as this will create

a harmonic currents
b eddy currents
c electrolytic action
d high inductance.

25. If a cable has a design current of 12.5 A and is 12 m long, what is its voltage drop if the volt drop per ampere per metre is 4.4 mV?

a 4.22 V
b 0.66 V
c 0.34 V
d 0.24 V.

26. If a 75 mm × 25 mm trunking has a factor of 738 and a stranded 2.5 mm^2 copper cable has a factor of 11.4, approximately how many cables can be installed in the trunking?

a 82
b 71
c 65
d 43.

27. All of the following are important factors for the effective termination of a multicore armoured PVC-insulated cable *except*

- **a** a need for access to inspect or provide maintenance
- **b** having knowledge of any high operating temperatures
- **c** the joint should be mechanically and electrically sound
- **d** having an understanding of the internal core connections.

28. To prevent cables from becoming damaged a conduit installation should be

- **a** wired in stages and then painted to protect it from corrosion
- **b** fitted with inspection type boxes before or after fabricated bends
- **c** erected with at least two outlet boxes for every 3 m length
- **d** completely erected before wiring commences.

29. A double set in a piece of conduit is equivalent to

- **a** two 90° bends
- **b** one 90° bend
- **c** two 45° bends
- **d** one 45° bend.

30. A **crampet** is a conduit accessory that is used to

- **a** fix conduit to a floor prior to the floor being covered with concrete
- **b** hold a length of conduit firmly in the jaws of a bending machine vice
- **c** support conduit and reduce vibration from the building structure
- **d** fix close lengths of conduit that run parallel between floors of a building.

31. What is the maximum length of span for a PVC cable having an HFOR sheath supported by a catenary wire?

- **a** no limit
- **b** 50 m
- **c** 20 m
- **d** 15 m.

32. Which one of the following is a situation where a cable tray would be a consideration?

- **a** high ambient temperature in a boiler room
- **b** long vertical surface drops in a church
- **c** pipework obstructing cable run in a factory
- **d** horizontal surface wiring in an office building.

33. Which one of the following flexible cords has an operating temperature of 185°C?

- **a** polychloroprene insulated
- **b** rubber-insulated
- **c** polyvinyl chloride-insulated
- **d** glass fibre-insulated.

34. On the length of trunking shown in Fig. 5, the distance *X* is twice the width of *Y*. This allows the trunking to be bent at

- **a** 90°
- **b** 75°
- **c** 45°
- **d** 30°.

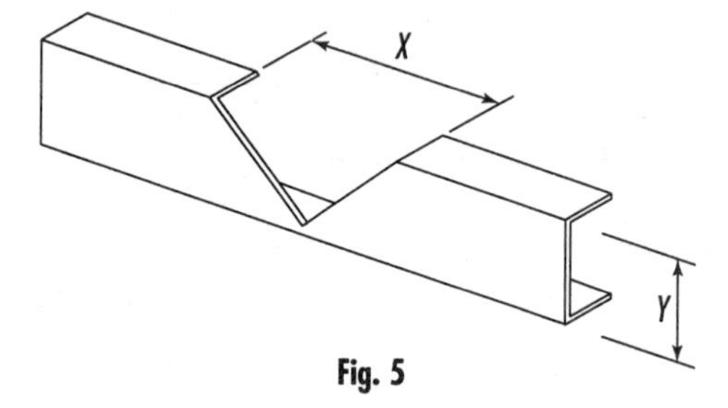

Fig. 5

35. It is possible to run a telecommunication circuit within the same enclosure as a low voltage final circuit provided the former is of the same

- **a** insulation voltage
- **b** cross-sectional area
- **c** supply frequency
- **d** cable manufacturer.

36. What is the name given to the wires attached to the busbars in Fig. 6?

- **a** sub-main cables
- **b** switch links
- **c** interconnectors
- **d** main tails.

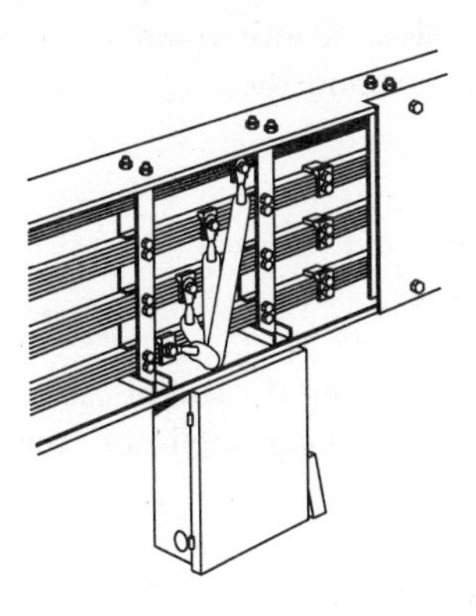

Fig. 6

37. In Fig. 6 above, the live conductors are normally connected from top to bottom in the following order:

a N, L1, L2 and L3
b L1, L2, L3 and N
c L2, N, L1 and L3
d L3, L2, L1 and N.

38. All of the following are methods to avoid the effects of electromagnetism when single-core armoured live cables pass through a metal enclosure, *except*

a use of only one entry hole
b making a slot in the entry plate
c using a non-ferrous entry plate
d connecting a discharge resistor in the circuit.

39. What is the name of the cable ladder accessory shown in Fig. 7?

a end section
b section joiner
c through piece
d left-hand reducer.

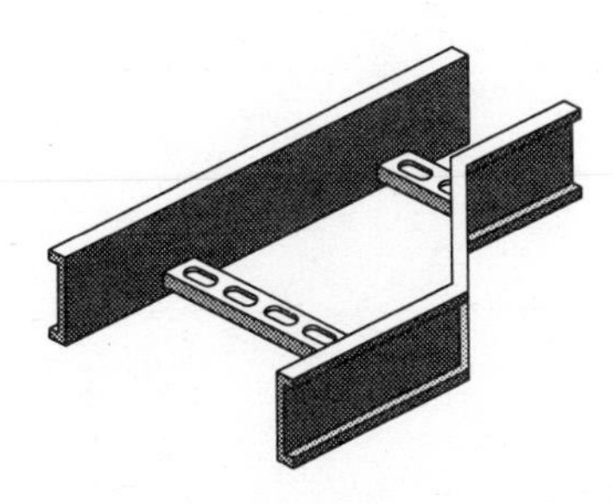

Fig. 7

40. The outer diameter of a stranded 10 mm^2 PVC-insulated single-core cable is 6.8 mm. What is its approximate cross-sectional area?

a 36 mm^2
b 22 mm^2
c 18 mm^2
d 12 mm^2.

41. In Question 40 above, what is the area of the cable's insulation?

a 36 mm^2
b 26 mm^2
c 13 mm^2
d 10 mm^2.

42. In a rising main wiring system, in what part of the installation would you be most likely to find a **fire barrier** installed?

a at an end section
b between two floors
c at the intake position
d along a lateral run.

43. The term **grouping factor** applies to all of the following cable dispositions *except*

a armoured cables bunched together on a cable tray
b MI cables clipped together, running horizontally along a wall
c PVC insulated, PVC sheathed cables passing through wooden joists
d PVC insulated, PVC sheathed cables tightly packed in trunking.

44. A certain cable has a design current of 8 A and voltage drop of 15 $mVA^{-1}m^{-1}$. If a maximum voltage drop of 4.0% is allowed, the longest length which can be used from a 240 V supply is

a 120 m
b 80 m
c 40 m
d 33 m.

45. In Table 54G of the *IEE Wiring Regulations* the minimum cross-sectional area of a protective conductor in relation to a phase conductor of 35 mm^2 is

a 16 mm^2
b 10 mm^2
c 6 mm^2
d 4 mm^2.

46. In Question 45 above, an alternative method of determining the size of protective conductor (S) is by calculation using the formula:

a $S = \sqrt{(I^2t)}/k$
b $S = I^2Rt$
c $S = V^2/Rt$
d $S = Pt$.

47. An extra low voltage system which is not electrically separated from earth but which satisfies all the requirements of a separated extra low voltage system is called

a SELV
b DELV
c FELV
d PELV.

48. The colour identification for a functional earthing conductor is

a cream
b green
c green and yellow
d orange.

49. When assessing the size of cable for a free-standing cooking appliance, consideration should be given to its

a load factor
b space factor
c diversity factor
d power factor.

50. The assessment of current demand for a cooking appliance is based on the first 10 A of total rated current plus 30% of the remainder current plus 5 A if the control point incorporates a 13 A socket outlet. What is the assessment for a 12 kW/240 V cooking appliance?

a 50 A
b 45 A
c 28 A
d 27 A.

ASSESSMENT 6.3

Answers, Hints and References

1. **d** Conductor resistance is inversely proportional to cross-sectional area.
2. **b** *See* manufacturers' catalogues.
3. **a.**
4. **a** *See* Ref. 2, p 32.
5. **d** It has four valence atoms and a negative temperature coefficient of resistance.
6. **c** This refers to a type of circuit protective device.
7. **c** See Ref. 4, Table 52B.
8. **a.**
9. **c** *See* manufacturers' catalogues.
10. **c** *See* manufacturers' catalogues.
11. **b.**
12. **a** *See* Ref. 19, p 17, Fig. 1.18.
13. **b** *See* Ref. 1, p 72.
14. **d** *See* Ref. 1, p 73.
15. **c.**
16. **b.**
17. **d.**
18. **c** *See* Ref. 18, Table 4C.
19. **a.**
20. **c.**
21. **d.**
22. **b** *See* Ref. 18, Tables 5A and 5B.
23. **d** *See* Ref. 4, Table 51 and Note 1.
24. **c** *See* Ref. 18, p 79.
25. **b** *See* Ref. 2, p 73, Step 5.
26. **c** *See* Ref. 18, Table 5F.
27. **a** *See* Ref. 4, regulation 526–04–01.
28. **d** *See* Ref. 4, regulation 522–08–02.
29. **b** *See* Ref. 29, no.1, p 29.
30. **a** *See* Ref. 21, Book 2, p 46.
31. **a** *See* Ref. 18, Table 4B.
32. **c.**
33. **d.**
34. **a** *See* Ref. 21, Book 2, p 77.
35. **a** *See* Ref. 4, regulation 528–01–08.
36. **c.**
37. **b.**
38. **d** *See* Ref. 4, regulation 521–02–01.
39. **d** *See* manufacturers' catalogues.
40. **a** $A = \pi d^2/4$.
41. **b** $36 - 10 = 26$.
42. **b.**
43. **c** *See* Ref. 26, p 75.
44. **b** $9.6/(8 \times 0.015) = 80$.
45. **a.**
46. **a** *See* Ref. 4, regulation 543–01–03.
47. **d** *See* Ref. 4, definitions.
48. **a** *See* Ref. 4, Table 51A.
49. **c** *See* Ref. 4, regulation 311–01–01.
50. **d** $10 + 12 + 5$.

7

PRINCIPLES OF CIRCUIT INSTALLATION

To tackle the assessments in Topic 7 you will need to know:

- about electrical equipment that is installed at the intake position on consumers' premises;
- how single-phase and three-phase supplies are balanced;
- how power is measured in ac and dc circuits;
- connections of an energy meter and be able to calculate unit consumption and cost of electricity;
- requirements for lighting circuits, types of lamp and associated equipment;
- maintenance and charging of cells and also the circuit wiring of simple alarm and call systems;
- requirements for BS1363 ring and radial final circuits and BS4343 industrial plugs and sockets;
- forms of heating circuit for cookers, water heaters and space heaters;
- requirements for single-phase and three-phase motors, their terminal connections.

DEFINITIONS

Balanced system – an arrangement of wiring single-phase and three-phase loads in an electrical installation so that each phase of the supply can carry approximately the same current. Cables and switchgear can then be economically chosen.

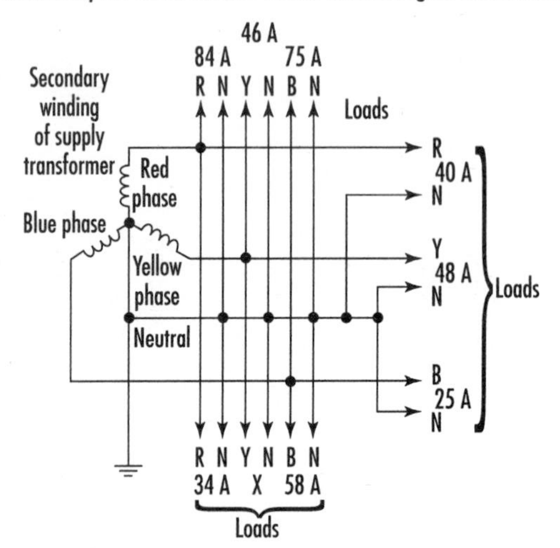

Load balancing

Capacitator – a component that has the ability to store electric charge. In dc circuits, current flow ceases when a capacitor becomes fully charged but in ac circuits the oscillating nature of the supply appears to make current continuously pass through the component.

Compact fluorescent lamp – a small type of fluorescent tube intended to replace the general lighting service lamp. It requires no starting control gear and may require a few seconds to ignite.

Contactor – a power controlling device that has an operating coil and contacts, used for opening and closing circuits.

Cut-out fuse – a cartridge fuse used by a Regional Electricity Company to protect a consumer's electrical installation. It is fitted in a cut-out at the intake position. For a domestic consumer, it is usually a Type 2 BS1361 fuse.

Delta connection – a three-phase connection in which each phase is joined to another phase so that it receives full supply voltage.

Emergency exit sign – an emergency sign that indicates to people in a building the route to escape (with an arrow) in the event of a fire.

Energy meter – an integrating meter that records the number of kilowatt hours of electricity used in a consumer's premises.

Energy regulator – a device used in an electrical cooking appliance to control the temperature of a heating element.

End-of-line resistor – a high ohmic value resistor placed at the end of triggering devices in an alarm system to keep a small current circulating around the circuit, keeping it self-monitored in case of a fault.

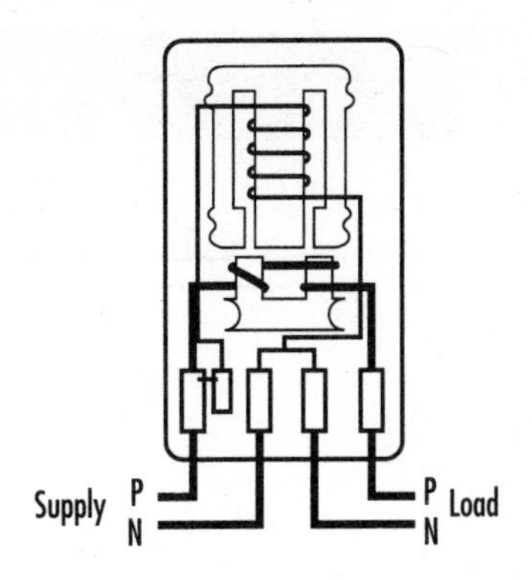

Energy meter

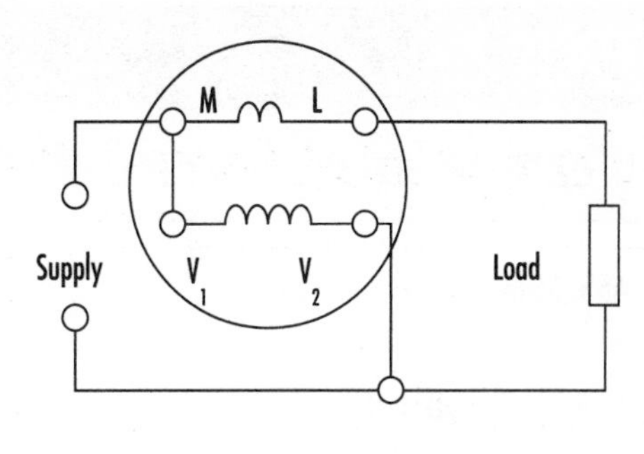

Wattmeter connection

Igniter – a component found in a discharge lamp circuit, often a high pressure sodium vapour (SON) lamp that is used to trigger a high voltage pulse to operate the lamp when it is first switched on.
Induction motor – a motor that operates by electromagnetic induction that can be designed for single-phase or three-phase working.
Integrating meter – see energy meter.
Lead–acid cell – a secondary rechargeable cell consisting of two lead plates immersed in an electrolyte (acid or diluted sulphuric acid) and capable of delivering a working voltage of 2 V.
Low-pressure sodium vapour lamp – a discharge lamp consisting of a U-shaped arc tube containing metallic sodium and an inert gas such as neon. When switched on it appears red and then slowly turns monochromatic yellow.
Normally closed alarm system – an alarm system whereby all the active operating contacts are in a closed position.
Normally open alarm system – an alarm system whereby all the active operating contacts are in an open position.
Passive infra-red detector – an intruder alarm detector unit that operates on movement of body heat across an invisible beam, often mounted in the corner of a room close to the ceiling giving a wide coverage of the room.

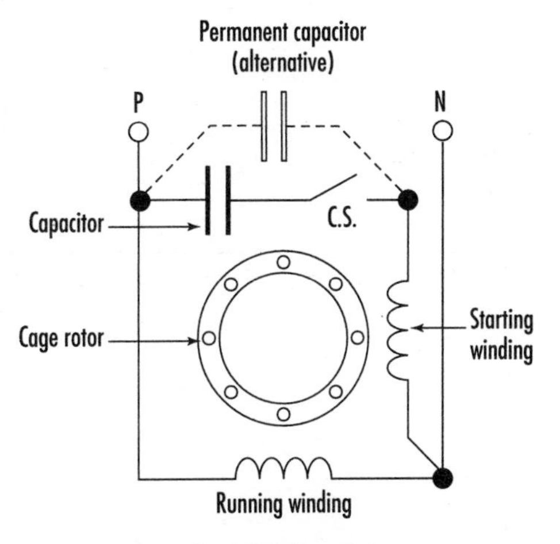

Wiring diagram of single-phase motor

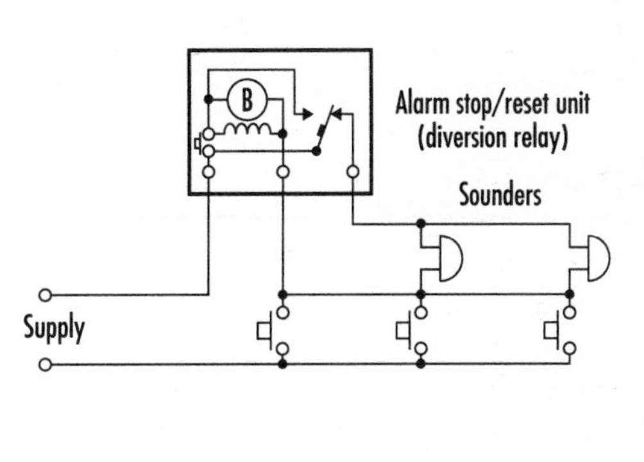

Fire alarm circuit

Rectifier unit – a device for changing ac into dc, often through solid-state semiconductor components such as pn diodes.
Space heater – a free standing or fixed electric heater with a resistive element for warming the interior of a room or building. Its control may be automatic and it may be provided with a fan to circulate warm air.
Star connections – a three-phase connection in which one end of each phase is joined to another phase and the other end connected to the supply.
Transformer – a static piece of electrical equipment mostly used for stepping-up/down voltage from one level to another.

TOPIC 7

Principles of Circuit Installation

ASSESSMENTS 7.1 – 7.3

Time allowed: 1 hour

Instructions

* You should have the following:
  Question Paper
  Answer Sheet
  HB pencil
  Metric ruler
* Enter your name and date at the top of the Answer Sheet.
* When you have decided a correct response to a question, on the Answer Sheet, draw a straight line across the appropriate letter using your HB pencil and ruler (see example below).
* If you make a mistake with your answer, change the original line into a cross and then repeat the previous instruction. There is only one answer to each question.
* Do not write on any page of the Question Paper.
* Make sure you read each question carefully and try to answer all the questions in the allotted time.

Example:

~~a~~ 400 V	~~a~~ (crossed) 400 V
b 315 V	**b** 315 V
c 230 V	~~c~~ 230 V
d 110 V	**d** 110 V

ASSESSMENT 7.1

1. What is the name of the electrical instrument shown in Fig. 1?

   a ac/dc wattmeter
   b integrating meter
   c consumption meter
   d power factor meter.

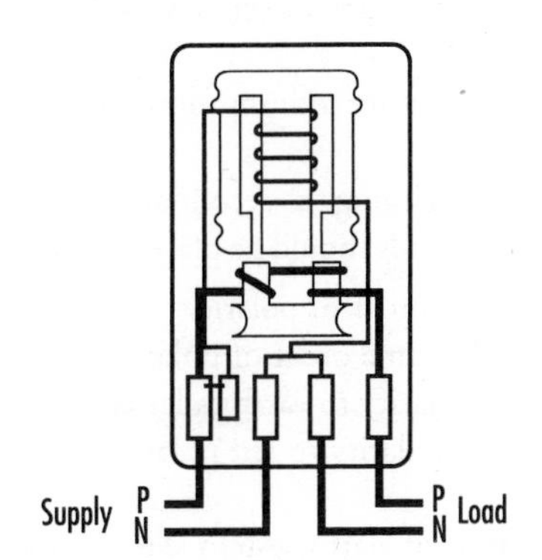

Fig. 1

2. The instrument in Fig. 1 is used for the purpose of

   a recording electrical energy
   b measuring consumed power
   c monitoring supply frequency
   d keeping supply voltage steady.

3. The instrument in Fig. 1 is commonly found at the intake position of a consumer's installation and is normally the responsibility of the

   a owner of the premises
   b regional electricity company
   c electrical subcontractor
   d property insurance company.

4. A 100 A, BS1361 Type 2 cartridge fuse is often found at the intake position of a domestic installation inside the

   a main switch
   b main fuseboard
   c cut out
   d kWh meter.

5. The past and present meter readings of a consumer's kilowatt hour energy meter are 54 750 and 55 000 respectively. If the cost of electricity is 7.31p/unit and a standing charge of £13.65 is to be added, what is the total cost?

   a £41.85
   b £18.80
   c £31.93
   d £20.88.

6. The instrument in Fig. 2 is called a

   a lightmeter
   b voltmeter
   c wattmeter
   d galvanometer.

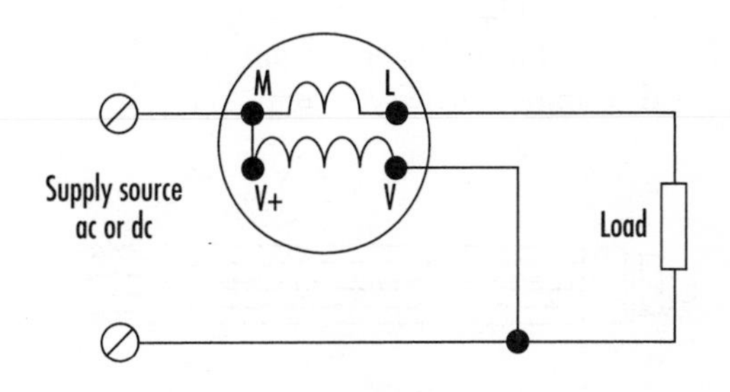

Fig. 2

7. What does the instrument in Fig. 2 measure?

- **a** temperature
- **b** power factor
- **c** energy
- **d** power.

8. In a single-phase ac circuit containing a resistive element a voltmeter measured a voltage of 240 V and an ammeter measured a current of 5 A. What power is consumed by the circuit?

- **a** 1.2 kW
- **b** 4.8 W
- **c** 20.8 mW
- **d** none.

9. In a two-wire dc circuit containing a resistive element, a wattmeter measures the power to be 3 kW. If the supply voltage is 240 V, what current would flow in the circuit?

- **a** 0.72 kA
- **b** 12.5 A
- **c** 8.0 mA
- **d** none.

10. When a three-phase, four-wire supply system is balanced

- **a** all phases rotate in the same direction as each other
- **b** all phases are designed to carry approximately the same current
- **c** no current will flow along the neutral conductor
- **d** the voltage between phases is the same as the voltage to earth.

11. The type of lamp shown in Fig. 3 is known as a

- **a** low pressure sodium lamp
- **b** general lighting service lamp
- **c** single-ended halogen lamp
- **d** compact fluorescent lamp.

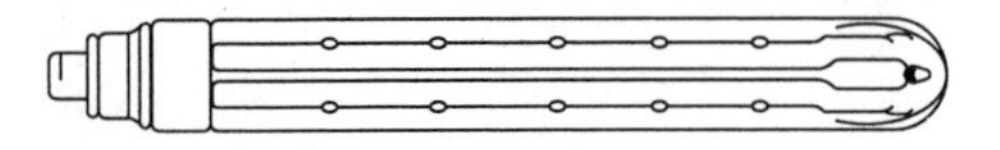

Fig. 3

12. Where a B22 lampholder is used in a bathroom and *not* fitted with a protected skirt, it should be

- **a** of a waterproof design
- **b** a semi-enclosed luminaire
- **c** 2.5 m away from the bath
- **d** inside a totally enclosed luminaire.

13. A lampholder for a filament lamp should not be installed in a final circuit operating at a voltage *exceeding*

- **a** 255 V
- **b** 250 V
- **c** 240 V
- **d** 235 V.

14. Which one of the following lamp circuits does *not* require control gear?

- **a** fluorescent tube circuit
- **b** tungsten halogen lamp circuit
- **c** high pressure mercury vapour lamp circuit
- **d** low pressure sodium vapour lamp circuit.

15. When charging a secondary battery, the charger's leads should be placed on the battery

- **a** in any order of polarity
- **b** in the same order of polarity
- **c** red polarity to earth polarity
- **d** black polarity to red polarity.

16. The type of alarm system shown in Fig. 4 is known as a

- **a** normally open circuit system
- **b** normally closed circuit system
- **c** manually-operated system
- **d** audio-visual indicator system.

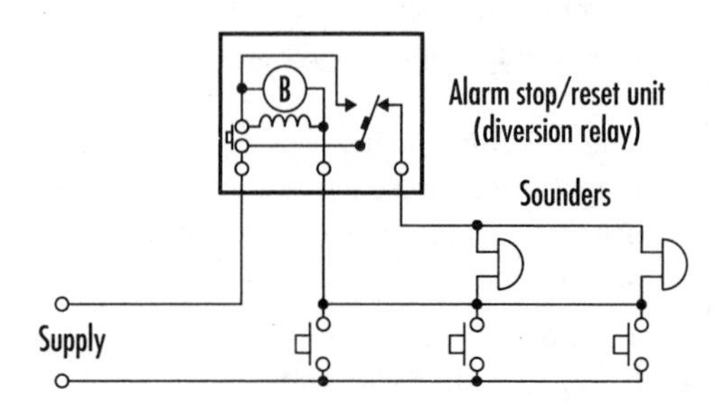

Fig. 4

17. In the discharge lamp circuit shown in Fig. 5, the component marked 'X' is called a

a transformer ballast
b pf correction capacitor
c starting resistor
d discharge arrestor.

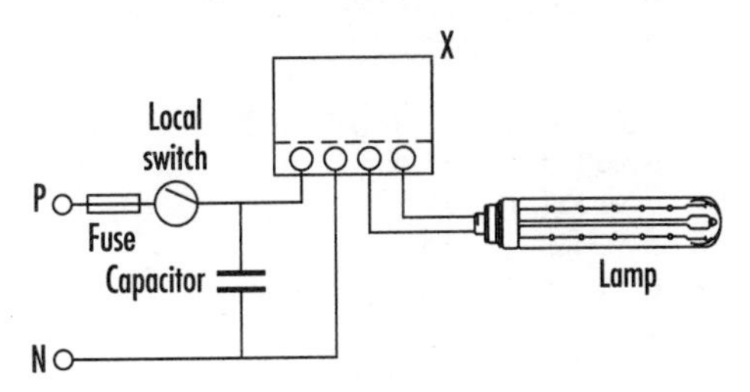

Fig. 5

18. Some discharge lamps have capacitors connected across the phase and neutral terminals. This is to

a discharge transient currents
b reduce lamp flicker
c improve circuit power factor
d reduce the stroboscopic effect.

19. In an intruder alarm circuit the device which is used to detect movement of body heat across an invisible beam is called a

a piezoelectric transducer
b ultrasound detector
c thermoelectric transducer
d passive infra-red detector.

20. In a simple **normally closed circuit** fire alarm system triggering devices are wired

a in parallel with each other
b in series with each other
c as a series/parallel combination
d as a ring main arrangement.

21. In a non-maintained emergency lighting system the emergency lights are

a illuminated after supply failure
b illuminated from a separate source
c repaired only during holiday periods
d off only during low battery voltage.

22. The maximum floor area that can be served with BS1363 socket outlets from a 32 A ring final circuit is

a 120 mm^2
b 100 mm^2
c 50 mm^2
d unlimited.

23. Non-fused spurs can be connected to a ring final circuit in all the following places *except* at a

a socket outlet not on the ring
b joint box wired in the ring
c consumer unit or fuseboard
d socket outlet on the ring.

24. What is the correct name of the device in Fig. 6 which is used to control the heat to a radiant ring on a cooker?

a thermal relay
b energy regulator
c thermostat
d control knob.

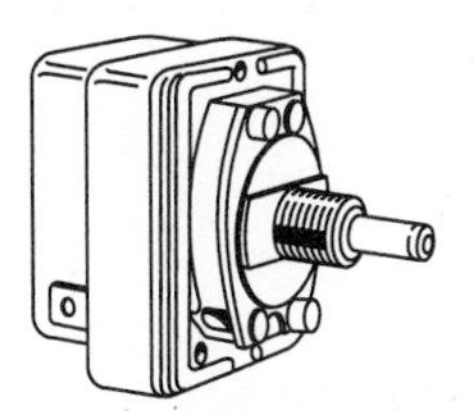

Fig. 6

25. The terminal markings for the running winding on the motor shown in Fig. 7 are

a A1 and A2
b B1 and B2
c U1 and U2
d Z1 and Z2.

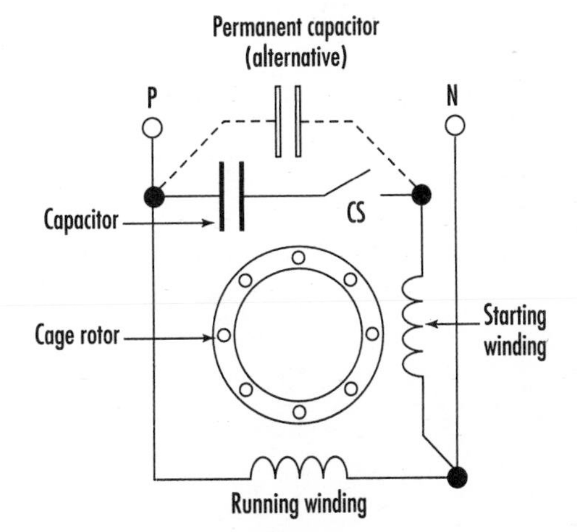

Fig. 7

ASSESSMENT 7.1

Answers, Hints and References

1. **b** *See* Ref. 1, pp 36–39.
2. **a** It does not measure power nor display frequency or keep the voltage steady.
3. **b** The REC's responsibility ends at the kWh meter.
4. **c** *See* Ref. 18, p 1.
5. **c** *See* Ref. 1, pp 36–39.
6. **c** *See* Ref. 1, p 46.
7. **d**.
8. **a** P = $V \times I$.
9. **b** I = P/V.
10. **c**.
11. **a** *See* Ref. 1, p 57, Fig. 3.28.
12. **d** *See* Ref. 4, regulation 601–11–01.
13. **b** *See* Ref. 4, regulation 553–03–02.
14. **b** *See* Ref. 1, pp 51–52.
15. **b** *See* Ref. 7, p 47, Fig. 2.30.
16. **a** *See* Ref. 1, p 62, Fig. 3.35(a).
17. **a** *See* Ref. 1, p 57, Fig. 3.29.
18. **c** *See* Ref. 1, p 55.
19. **d** *See* Ref. 1, p 65.
20. **b** *See* Ref. 1, p 62, Fig. 3.35(b).
21. **a** *See* Ref. 12, p 60.
22. **b** *See* Ref. 18, p 124.
23. **a** *See* Ref. 18, p 125.
24. **b** *See* Ref. 7, p 114, Fig. 6.8.
25. **c** *See* Ref. 7, p 75, Fig. 4.7.

ASSESSMENT 7.2

1. All the following are likely to be installed at the intake position of a consumer's electrical installation, *except* a

   a changeover switch
   b residual current device
   c kWh energy meter
   d time switch.

2. Which one of the following overcurrent circuit protective devices is used as a **cut-out fuse** in a single-phase, domestic dwelling?

   a 125 A, BS88, cartridge fuse
   b 100 A, BS1361, cartridge fuse
   c 80 A, BS3036, semi-enclosed fuse
   d 60 A, BS3871, M9 miniature circuit breaker.

3. The meter tails in Fig. 1 will normally be connected to the electricity supply authority's

   a main busbar
   b main switch
   c energy meter
   d cut-out fuse.

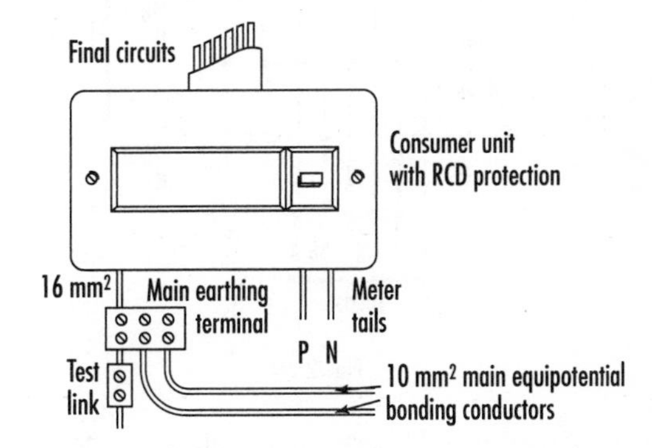

Fig. 1

4. As a general rule a double-wound, step-down transformer has the ac supply connections made to its

   a secondary winding
   b primary winding
   c delta winding
   d star winding.

5. How many units of electricity are used by a 7.5 kW shower heater kept on for 5 minutes?

   a 0.625
   b 0.455
   c 0.125
   d 0.037.

6. For the system shown in Fig. 2 to be ideally balanced, the load marked X must read

a 94 A
b 64 A
c 58 A
d 34 A

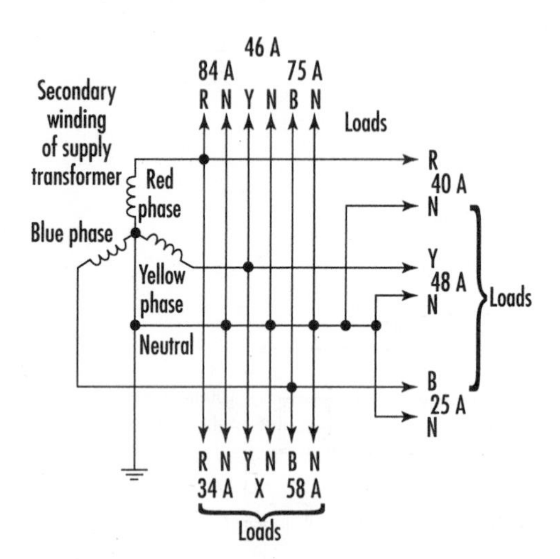

Fig. 2

7. A consumer's electrical installation is supplied with a single-phase supply comprising a

a fuse in both live conductors
b cut-out fuse and neutral link
c high resistance earth electrode
d fixed speed kilowatt hour meter.

8. Which one of the following is designed to produce an electromotive force (emf) of 2 V per cell?

a dry battery
b alkaline
c lead–acid
d Leclanché.

9. Which one of the wattmeter connections shown in Fig. 3 is wired correctly?

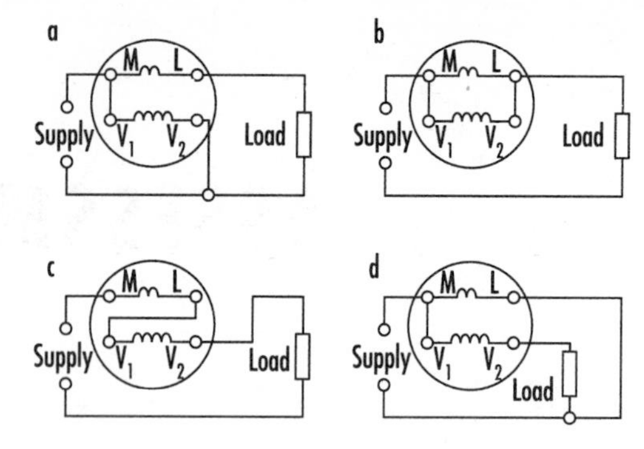

Fig. 3

10. When a B15 bayonet lampholder is installed in a lighting circuit, the maximum rating of the overcurrent protective device must *not* exceed

a 30 A
b 16 A
c 10 A
d 6 A.

11. Which one of the following is acceptable for a non-metallic luminaire installed within 2.5 m of a bath?

a it must not have a lamp wattage exceeding 100 W
b it must be of shatterproof construction
c it must be semi-enclosed and have drain holes
d it must be totally enclosed or have a protective skirt.

12. Fig. 4 shows an emergency exit sign. This should be installed in all the following positions, *except*

a places where there is a potential hazard
b along corridors and intersections
c on staircases and changes of floor level
d inside toilets and washrooms.

Fig.4

13. The operating contacts of a single-pole thermostat controlling an immersion heater should be connected

a between phase and neutral
b in the phase conductor
c in the neutral conductor
d between phase and earth.

14. The type of lamp shown in Fig. 5 is called a

a high pressure discharge lamp
b general lighting service lamp
c single-ended halogen lamp
d compact fluorescent lamp.

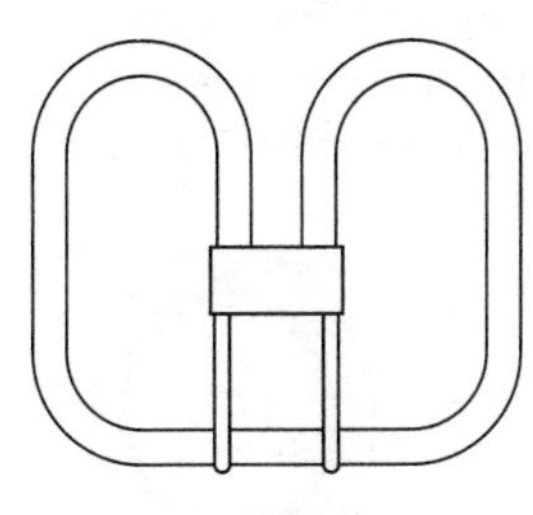

Fig. 5

15. When initially switched on, the light emitted by a low pressure sodium vapour lamp is

a blue
b red
c white
d violet.

16. Which arrangement in Fig. 6 shows the correct balancing of a three single-phase ac supply?

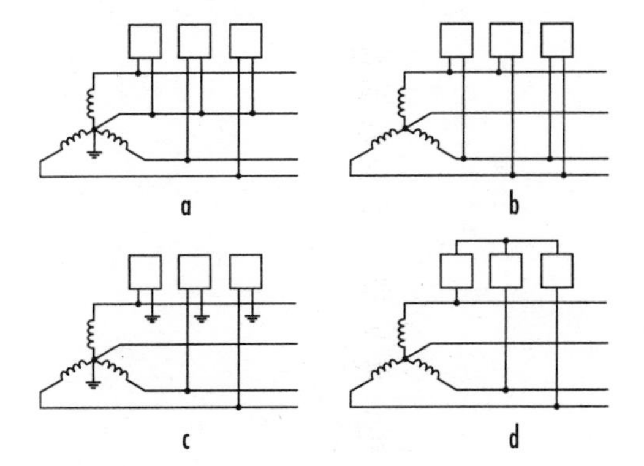

Fig. 6

17. In a maintained emergency lighting system, the lights are

a illuminated only after supply failure
b in operation at all material times
c regularly repaired every 3 months
d off only during low battery voltage.

18. In the battery charging circuit shown in Fig. 7 the purpose of the transformer is to

a reduce circuit voltage to 12 V
b protect the battery and ammeter
c change the ac supply into dc
d trickle charge the battery.

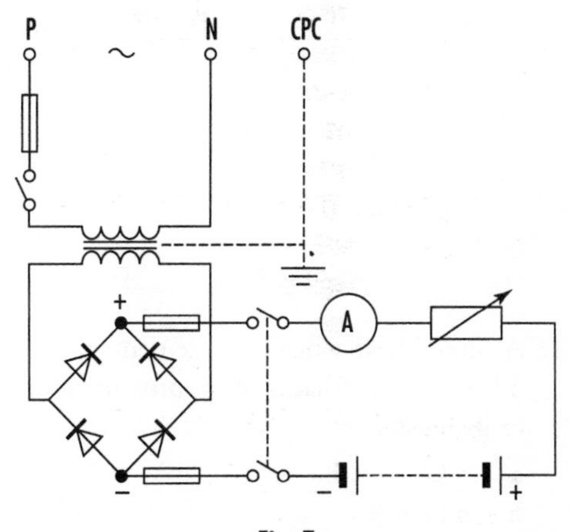

Fig. 7

19. In a closed-circuit intruder alarm circuit, cutting a wire feeding a window contact would

a cause the sounder to operate
b remove its zone of protection
c make the sounder inoperative
d rupture the protective fuse.

20. When assessing current demand of electrical equipment, a lighting outlet for a filament lamp is to be at least

a 5.0 A
b 2.5 A
c 1.0 A
d 0.5 A.

21. The current ratings of the fuses shown in Fig. 8 from left to right are normally

a 30 A, 32 A, 15 A, 10 A
b 45 A, 32 A, 12 A, 6 A
c 30 A, 30 A, 10 A, 6 A
d 45 A, 30 A, 15 A, 5 A.

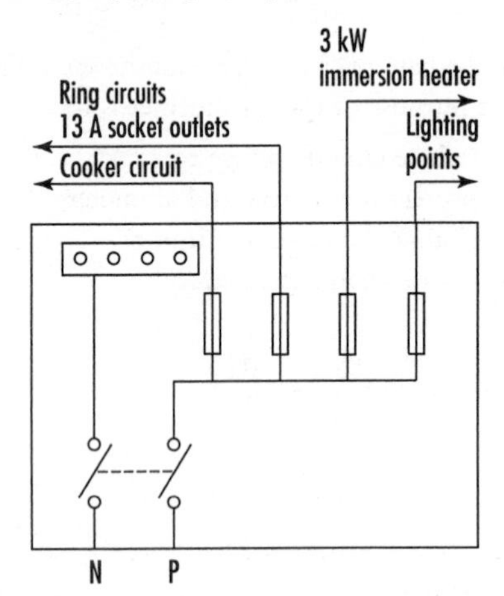

Fig. 8

22. A space heater has an element rated at 3 kW/240 V. What is the approximate value of its resistance?

a 80 Ω
b 25 Ω
c 20 Ω
d 13 Ω.

23. When PVC insulated cable is used to wire BS1363 socket outlets from a 20 A overcurrent protective device, the minimum size cable that can be used in a radial circuit is

a 6.0 mm^2
b 4.0 mm^2
c 2.5 mm^2
d 1.5 mm^2.

24. The BS4343 (BS EN 60309-2) industrial socket shown in Fig. 9 is identified by the colour red and is to be used only on ac supplies of

a 750 V
b 400 V
c 230 V
d 110 V.

25. If one of the circuit fuses protecting a three-phase, cage induction motor were not in place, the indication on starting the motor would be that it would

a not run and remain silent
b not run but hum noisily
c oscillate and try to run
d run at one-third of its speed.

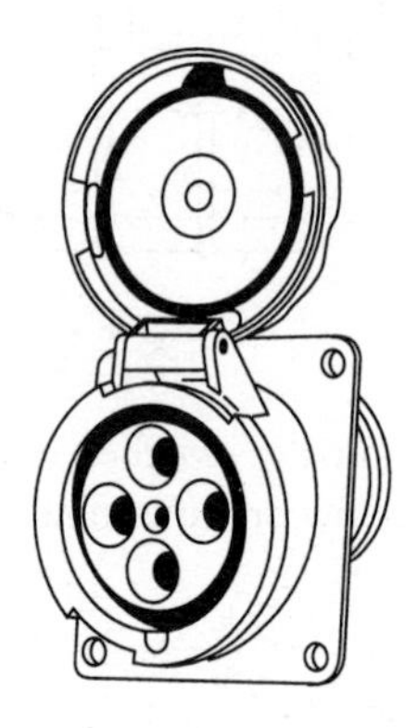

Fig. 9

ASSESSMENT 7.2

Answers, Hints and References

1. **a**.
2. **b** *See* Ref. 18, p 1.
3. **c** *See* Ref. 18, p 5.
4. **b** *See* Ref. 1, p 37, Fig. 3.1.
5. **a** 7.5 × 5/60.
6. **b** *See* Ref. 1, pp 40–44.
7. **b** *See* Ref. 18, p 1.
8. **c** *See* Ref. 7, pp 44–45.
9. **a** *See* Ref. 1, p 46, Fig. 3.12.
10. **d** *See* Ref. 4, Table 55B.
11. **d** *See* Ref. 4, regulation 601–11–01.
12. **d** *See* Ref. 12, p 97.
13. **b** *See* Ref. 7, p 115, Fig. 6.10.
14. **d** *See* Ref. 1, p 54, Fig. 3.23.
15. **b** *See* Ref. 1, p 57.
16. **a** *See* Ref. 1, pp 43–44.
17. **b** *See* Ref. 1, p 60.
18. **a** *See* Ref. 7, pp 47, Fig. 2.30.
19. **a** *See* Ref. 1, pp 63–65.
20. **d** *See* Ref. 18, Table 1A.
21. **d**.
22. **c** $R = V^2/P = 19.2\ \Omega$.
23. **c** *See* Ref. 18, Table 8A.
24. **b** *See* Ref. 1, p 49, Fig. 3.15.
25. **a**.

ASSESSMENT 7.3

1. A Regional Electricity Company may install the switch marked 'X' in Fig. 1 to permit the supply to the consumer to be disconnected

a without withdrawal of the cut-out fuse
b where shared occupancy of the premises exists
c if there is danger from a gas leak
d if the supply voltage rises above 10%.

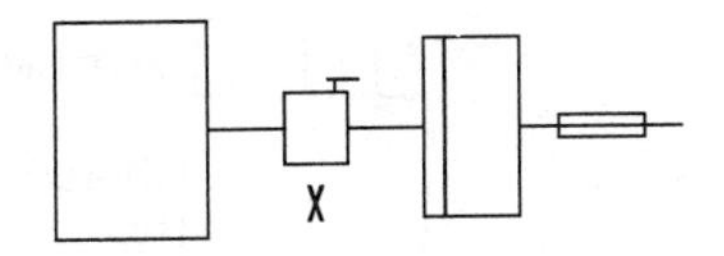

Fig. 1

2. A 400 V/230 V, three-phase, four-wire ac supply system is usually derived from a transformer with windings that are normally connected in

a delta primary and delta secondary
b delta primary and star secondary
c star primary and delta secondary
d star primary and star secondary.

3. What is the name of the component marked 'X' in Fig. 2 that will control the supply of electricity to 'off peak' storage heaters?

a circuit breaker
b relay unit
c changeover switch
d contactor.

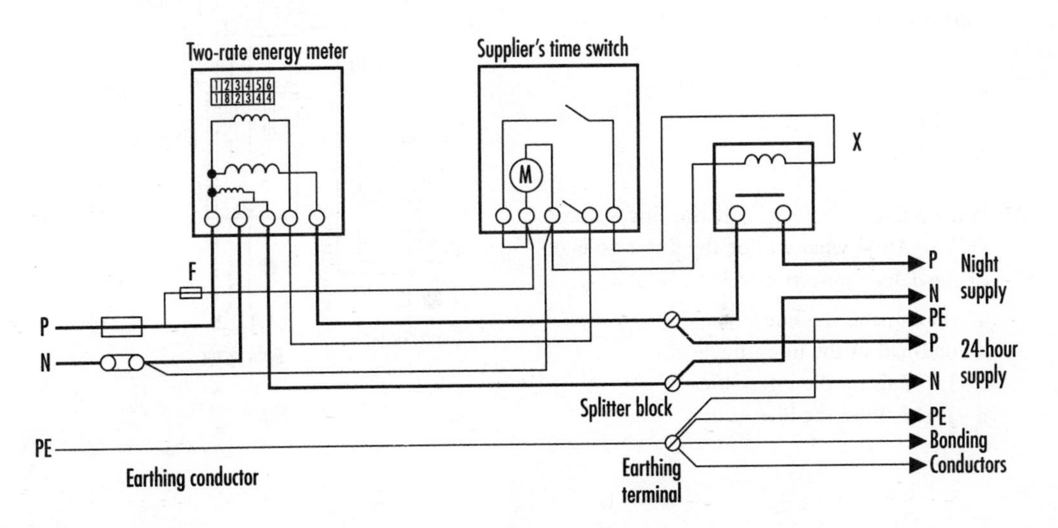

Fig. 2

4. When a three-phase, four-wire ac supply system is ideally balanced it means that

   a there is no current flowing in the neutral conductor
   b phase currents and line currents are equal
   c line currents are larger than phase currents
   d the neutral carries three times less current than the phases.

5. The 'normal' past and present meter readings of a consumer's kilowatt hour meter are 39 775 and 40 584 respectively at 7.31 pence per unit. If the 'low' readings were 5000 and 5123 at 2.71 pence per unit and a standing charge of £10.40 is to be added, what is the total bill exclusive of VAT?

   a £113.80
   b £103.40
   c £69.54
   d £72.87.

6. In Question 5 above, if an 8% VAT charge was added to the bill it would become

   a £122.90
   b £111.67
   c £75.10
   d £78.70.

7. In Fig. 3, if the line voltage (V_L) = 415 V, what will be the phase voltage of the **star connection**?

   a 250 V
   b 240 V
   c 230 V
   d 220 V.

8. With reference to Fig. 3 if the line voltage (V_L) = 415 V, what will be the phase voltage of the delta connection?

   a the same as the line value
   b one-half of the line value
   c 1.732 times the line value
   d 1.414 times the line value.

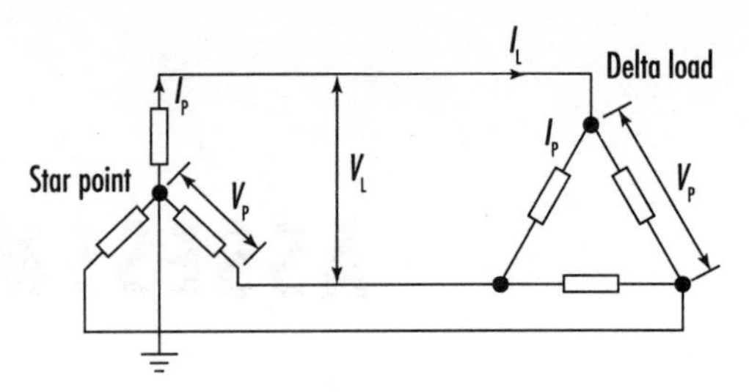

Fig. 3

9. Two-way and intermediate switching is most commonly used to control a lighting point

   a on three floor levels
   b around an L-shaped corridor
   c in a room with two doors
   d in three adjacent rooms.

10. Fig. 4 shows a three-phase and neutral supply to a heating circuit. At what point on the diagram would you connect a room thermostat?

   a 4
   b 3
   c 2
   d 1.

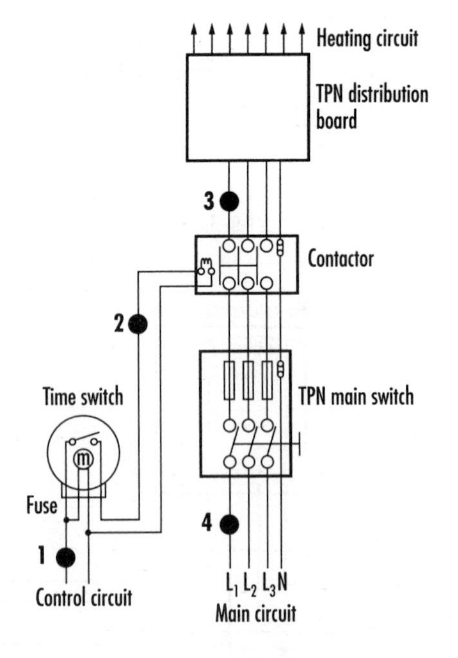

Fig. 4

11. Ignoring temperature change of the element, when a 3 kW/240 V immersion heater is supplied at half the normal voltage, its power is reduced by

a 75%
b 50%
c 25%
d 10%.

12. The control circuit for a low-pressure fluorescent discharge lamp will have all the following components *except*

a an igniter
b a starter switch
c ballast
d a pf capacitor.

13. When determining the loading of fluorescent tube circuits it is recommended that the voltamperes be determined by multiplying the rated lamp power by a factor of

a 3.0
b 2.5
c 1.8
d 0.8.

14. Why is it necessary to balance loads in a three-phase, four-wire ac supply system?

a the supply transformer and switchgear cannot become overloaded
b the electricity supply company can provide an efficient earthing system
c cables, equipment and switchgear can be chosen economically
d all single-phase circuits can be operated safely and independently.

15. For the transformer shown in Fig. 5 the secondary voltage (V_s) is

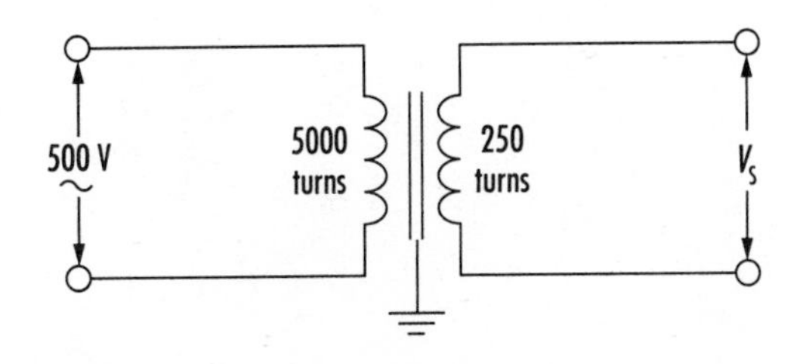

Fig. 5

a 10 kV
b 2.5 kV
c 25 V
d 5 V.

16. Which one of the following circuits in Fig. 6 is correct for connecting the ammeter and voltmeter instruments?

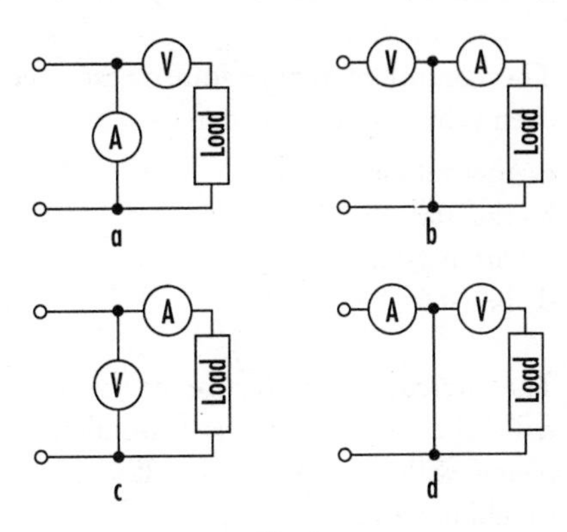

Fig. 6

17. If the load is resistive in Fig. 6 the correct circuit can be used to determine

a power
b energy
c capacitance
d reactance.

18. One method of monitoring a fire alarm system for faults to be traced is to incorporate in its circuit

a polarized sounders
b end-of-line resistors
c a power factor relay
d a diversion relay.

19. In a battery charging circuit the purpose of a **rectifier unit** is to

a reduce the voltage to 12 V
b protect the battery and ammeter
c change ac into dc
d trickle charge the battery.

20. A lead–acid secondary battery is being discharged at the rate of 3 A for 15 hours. If its efficiency is 90% what is the ampere hour charge required?

a 50 Ah
b 45 Ah
c 40.5 Ah
d 16.6 Ah.

21. The control of temperature in an electric oven is by a device called a

a thermostat
b rheostat
c simmerstat
d heatstat.

22. The overcurrent protective device in final radial circuits using 16 A socket-outlets (complying with BS4343 or BS EN 60309-2) should not exceed

a 45 A
b 32 A
c 30 A
d 20 A.

23. When general purpose BS1363 socket-outlets are being wired in a 30 A ring final circuit, each socket-outlet of a twin or multiple socket-outlet unit is to be regarded as

a one outlet
b two outlets
c the number per outlet
d a maximum of four outlets.

24. When a star-delta starter is used to start a three-phase motor, how many circuit conductors are required between the starter and motor?

a 6
b 4
c 3
d 2.

25. A motor with terminal markings identified as: U1–U2, V1–V2 and W1–W2 is likely to be a

a one-phase, split-phase motor
b three-phase ac motor
c dc compound motor
d dc shunt motor.

ASSESSMENT 7.3

Answers, Hints and References

1. **a**.
2. **b** *See* Ref. 1, p 37, Fig. 3.1.
3. **d** *See* Ref. 1, p 53, Fig. 4.4(c).
4. **a** *See* Ref. 1, p 44.
5. **d** £59.14 + £3.33 + £10.40.
6. **d**.
7. **b** *See* Ref. 12, pp 44–45.
8. **a** *See* Ref. 1, pp 44–45.
9. **a**.
10. **c** It has to come after the time clock.
11. **c** $(V/2)^2 \times 1/R$= ¼ Power.
12. **a** An igniter may be used to start a SON lamp.
13. **c** *See* Ref. 18, Table 1A, note.
14. **c**.
15. **c** *See* Ref. 7, p 90, Fig. 5.3.
16. **c**.
17. **a** *See* Ref. 7, p 34.
18. **b** *See* Ref. 1, p 62, Fig. 3.36.
19. **c** *See* Ref. 7, p 47, Fig. 2.30.
20. **a** *See* Ref. 7, p 50.
21. **c**.
22. **d** *See* Ref. 18, p 126.
23. **a** *See* Ref. 18, p 123.
24. **a** *See* Ref. 8, p 57, Fig. 3.24.
25. **b** *See* Ref. 8, p 57, Fig. 3.24.

8

THE INSTALLATION OF EARTHING EQUIPMENT

To tackle the assessments in Topic 8 you will need to know:

- the purpose of earthing the exposed conductive parts and extraneous conductive parts of an electrical installation;
- the operational requirements of protective equipment such as overcurrent devices and residual current devices;
- the purpose, forms and sizing of circuit protective conductors;
- how to select earthing conductors and earth electrodes;
- the purpose and means of providing main equipotential and supplementary bonding;
- the methods of protection for outlets and equipment used outside the equipotential zone.

DEFINITIONS

Bonding conductor – a protective conductor that provides equipotential bonding between a consumer's earthing arrangements and metallic pipework or structure not directly associated with the consumer's earthing (*see* main equipotential bonding and supplementary bonding below).
Circuit protective conductor – a protective conductor connecting exposed conductive parts of equipment to the main earthing terminal.
Circuit protective device – a device placed in an electrical circuit to provide protection against overcurrent (*see* fuses and miniature circuit breakers below).

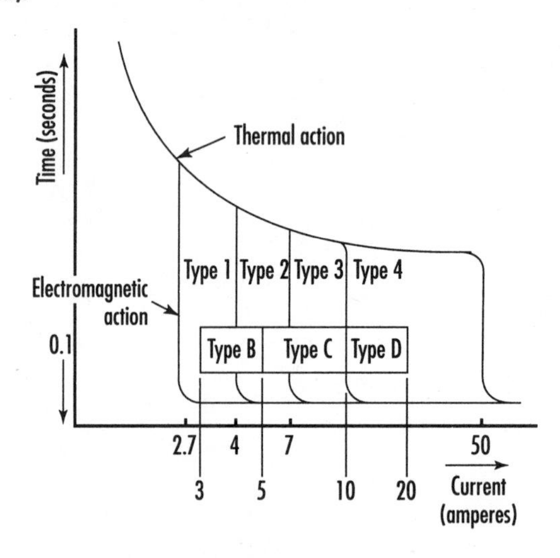

Time – current characteristics of miniature circuit breakers

Conductive part – metalwork and other material likely to conduct an electric current (*see* exposed and extraneous conductive parts below).
Direct contact – contact with live parts.
Disconnection time – the time taken for a protective device to disconnect the circuit it controls.
Earth clamp – a metal clip designed to fasten securely onto metal pipes to provide earth bonding (designed to BS951).
Earth electrode – a conductor or group of conductors effectively embedded in the general mass of earth, designed to discharge fault current into the mass of earth.

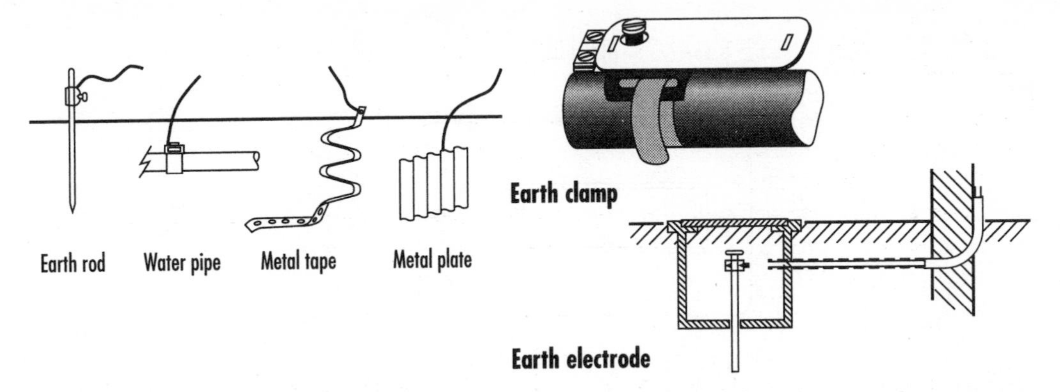

Earth fault – a fault current escaping to the general mass of earth often as a result of a live conductor touching earthed metalwork.
Earth fault loop impedance – the impedance (in ohms) of the path taken by an earth fault current starting and ending at the point of the fault.
Earthed – a connection made to the general mass of earth.

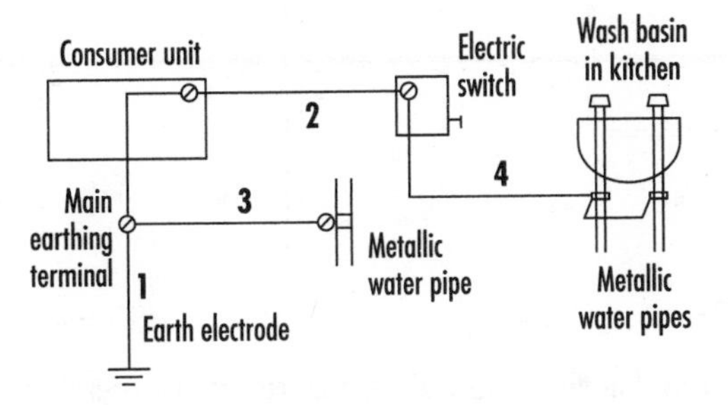

Earthing and bonding

Earthing conductor – a protective conductor connecting the main earthing terminal of a consumer's installation to an earth electrode or other means of earthing.
Earthing terminal (main) – a terminal block placed at the intake position of a consumer's electrical installation for the connection of protective conductors and bonding conductors.
Earthing system – an arrangement of the energy supply source to an installation and the method used for earthing the source (*see* Ref. 4, Types of Earthing System).
Electrical equipment – any item of electrical equipment associated with an electrical installation (Class I equipment requires to be earthed, Class II equipment does not require to be earthed, Class III equipment is supplied by SELV).
Equipotential bonding – a method of connecting exposed conductive parts and extraneous conductive parts at the same potential to avoid the risk of electric shock.
Exposed conductive part – a part of equipment which can be touched and although not a live part, may become live as a result of a fault.
Extraneous conductive part – a part liable to introduce a potential (generally earth potential) which is not part of the electrical installation.
Fuse – a circuit protective device for protecting a circuit from overcurrent. Common fuses are rewirable fuses (BS3036), cartridge fuses, (BS1361, BS1362 and BS88).
Fuse discrimination – a design arrangement for a fuse or other form of protective device nearest a fault to disconnect the faulty circuit before any upstream protective devices operate.
Fusing factor – a ratio of the rated minimum fusing current and the current rating of a circuit protective device.
Indirect contact – contact with exposed conductive parts that have become live under fault conditions.
Main equipotential bonding – an arrangement to bond together gas, water and other metalwork to the consumer's main earthing terminal.

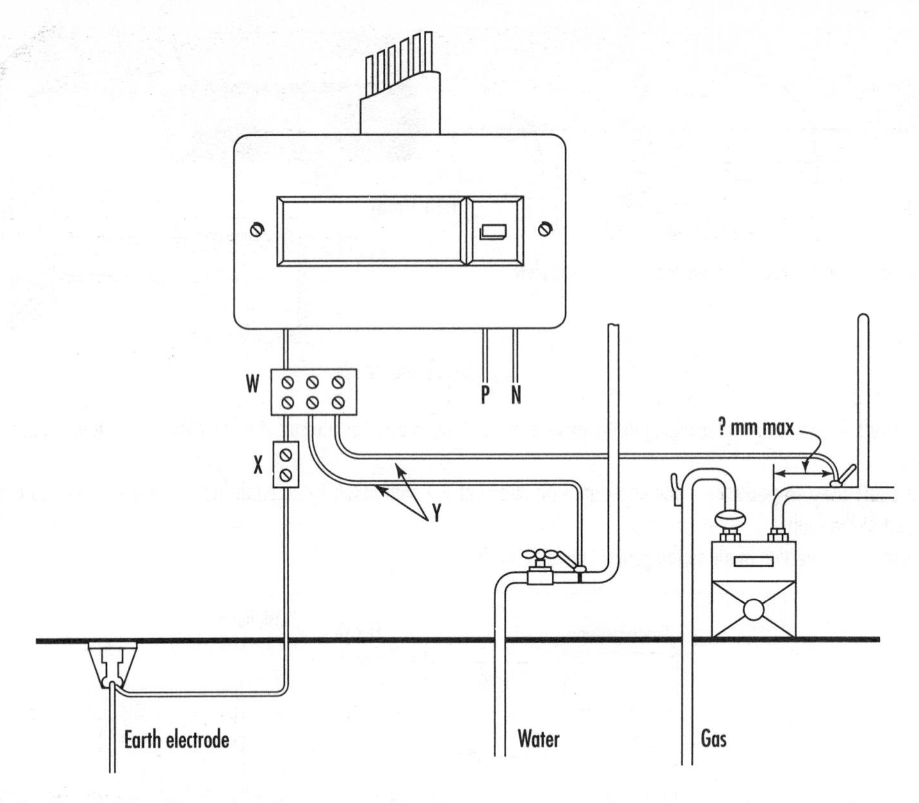

Main equipotential bonding

Material factor – a term expressed as the symbol *k* which takes into account resistivity, temperature coefficient and heat capacity of a conductor material.
Miniature circuit breaker (mcb) – a circuit protective device capable of detecting small overloads and short-circuit faults.
Overcurrent – a current exceeding the rated value. In an electrical circuit which is sound, it could be an overload.
PEN conductor – a conductor combining the functions of both a protective conductor and neutral conductor.
PME – an abbreviation for protective multiple earthing but is now known as a TN-C-S earthing system.
Residual current device – an earth leakage circuit breaker intended to be selected as a supplementary form of protection against electric shock as well as risk of fire.

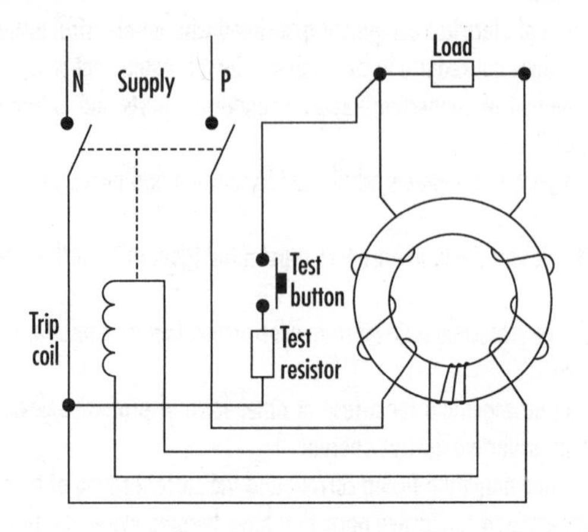

Residential current device

Supplementary equipotential bonding – the bonding of extraneous conductive parts and exposed conductive parts.

TOPIC 8

The Installation of Earthing Equipment

ASSESSMENTS 8.1 – 8.3

Time allowed: 1 hour

Instructions

* You should have the following:

  Question Paper
  Answer Sheet
  HB pencil
  Metric ruler

* Enter your name and date at the top of the Answer Sheet.
* When you have decided a correct response to a question, on the Answer Sheet, draw a straight line across the appropriate letter using your HB pencil and ruler (see example below).
* If you make a mistake with your answer, change the original line into a cross and then repeat the previous instruction. There is only one answer to each question.
* Do not write on any page of the Question Paper.
* Make sure you read each question carefully and try to answer all the questions in the allotted time.

  Example:

~~a~~	400 V	~~a~~ (crossed)	400 V
b	315 V	**b**	315 V
c	230 V	~~c~~	230 V
d	110 V	**d**	110 V

ASSESSMENT 8.1

1. An electrical installation is earthed so that

- **a** the risk of electric shock is minimized
- **b** short-circuit faults can be made safe
- **c** residual current devices can operate
- **d** the supply voltage will remain constant.

2. With reference to Fig. 1 the item marked W is called a

- **a** test link
- **b** main earthing terminal
- **c** PME connection
- **d** RCD test point.

3. With reference to Fig. 1 the conductors marked Y are called

- **a** supplementary bonding conductors
- **b** main equipotential bonding conductors
- **c** main earthing conductors
- **d** circuit protective conductors.

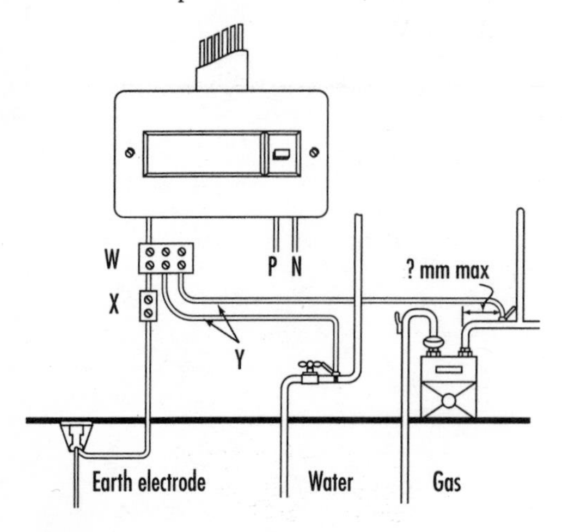

Fig. 1

4. In Fig. 1 the **bonding** connection for the gas service should be within

- **a** 600 mm of the gas meter
- **b** 500 mm of the gas meter
- **c** 300 mm of the gas meter
- **d** 100 mm of the gas meter.

5. The type of earthing system shown in Fig. 1 is called a

- **a** TN-S system
- **b** IT system
- **c** TT system
- **d** TN-C-S system.

6. An equipotential zone in a consumer's electrical installation is carried by

- **a** earthing the main water pipe
- **b** installing maximum earth points
- **c** using a copper earth electrode
- **d** earthing and bonding all metalwork.

7. The opposition to current flow when a final circuit protective device trips out or ruptures, as a result of an earth fault, is called

- **a** resistance
- **b** impedance
- **c** inductance
- **d** reactance.

8. All of the following are recognized conductive parts of electrical wiring systems, *except*

- **a** rigid metal trunking
- **b** pliable metal conduit
- **c** metal covering of cables
- **d** circuit protective conductors.

9. What is the name given to the metal part that could introduce a potential into an electrical installation from outside?

a equipotential conductive part
b extraneous conductive part
c supplementary conductive part
d isolated conductive part.

10. Which one of the numbered conductors in Fig. 2 is a **circuit protective conductor**?

a 4
b 3
c 2
d 1.

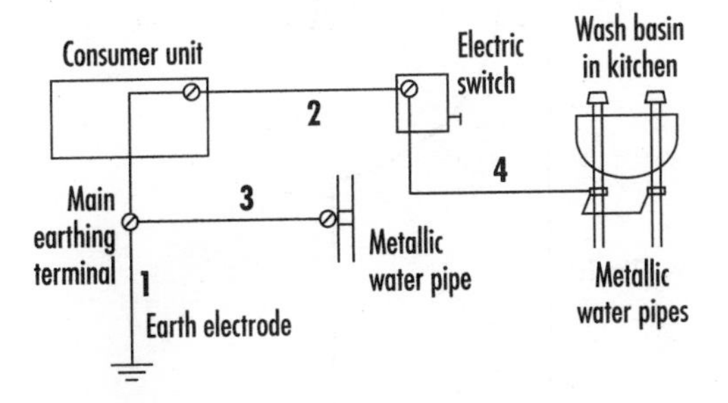

Fig. 2

11. Which one of the numbered conductors in Fig. 2 is a **supplementary equipotential bonding** conductor?

a 4
b 3
c 2
d 1.

12. Which one of the numbered conductors in Fig. 2 is an **earthing conductor**?

a 4
b 3
c 2
d 1.

13. When all exposed metalwork of an electrical installation is earthed and bonded together, the protective measure should cause earth faults to

a discharge outside consumers' premises
b dissipate safely through the wiring
c be limited to low levels of current
d automatically disconnect final circuits.

14. The *IEE Wiring Regulations* exempt from earthing all of the following items of metalwork *except*

a brackets and metal parts of overhead wiring
b fixing screws for non–metallic accessories
c accessible structural steelwork of building
d inaccessible conduit not longer than 150 mm in length.

15. Which one of the following in Fig. 3 is *not* recognized as an earth electrode?

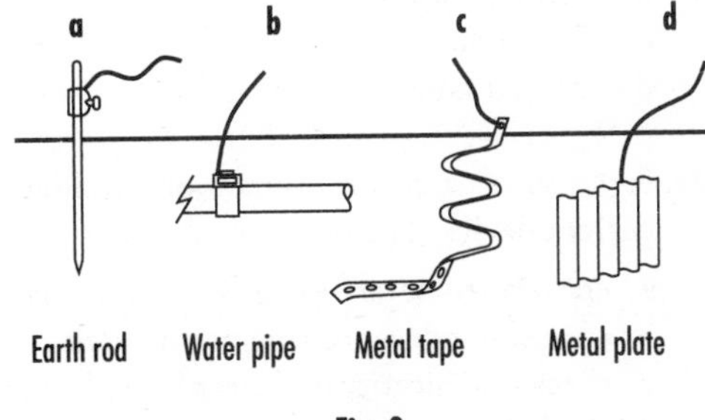

Fig. 3

16. A **residual current device** is a circuit protective device which operates when

a fuses and mcbs fail to function
b an overload or short circuit occurs
c a known level of earth leakage occurs
d the supply voltage falls below 240 V.

17. Which one of the following shall *not* be selected as a protective conductor?

a insulated cable conductor
b armouring of a cable
c metal gas service pipe
d steel conduit or trunking.

18. The words inscribed on the BS951 earth clamp shown in Fig. 4 should read

a Electrical Earth Connection – Do Not Move
b Earth Connection – Do Not Move
c Safety Electrical Earth – Do Not Remove
d Connection to Earth – Do Not Remove.

Fig. 4

19. Which one of the following letter symbols is used in the *IEE Wiring Regulations* to denote a protective conductor's cross-sectional area?

a U
b C
c S
d T.

20. Consumer units and distribution fuseboards contain circuit protective devices for all the following reasons *except*

a open circuit faults
b short circuit faults
c earth faults
d overload faults.

21. Fig. 5 shows a circuit diagram of a **residual current device**. The test button is used to

a create a temporary earth fault on no load
b create an imbalance core magnetic field
c check the integrity of the earth conductor
d check the continuity of the live conductors.

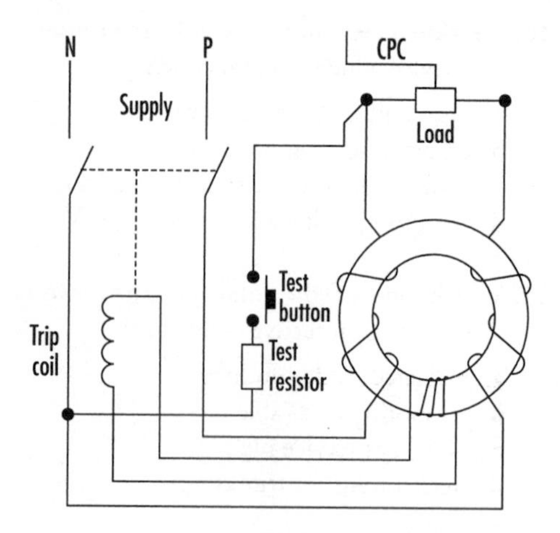

Fig. 5

22. The term **fusing factor** is concerned with the ratio of a fuse's

a short-circuit performance and its overload
b correction factor and working tolerance
c overload current and circuit power factor
d rupturing current and its rated current.

23. A miniature circuit breaker has two operating functions: one is to provide overload protection, the other is to provide

a short-circuit protection
b open-circuit protection
c small-leakage protection
d high-resistance protection.

24. Fig. 6 shows the operational characteristics of a circuit protective device known as a

a semi-enclosed rewirable fuse
b miniature circuit breaker
c high breaking capacity fuse
d residual current device.

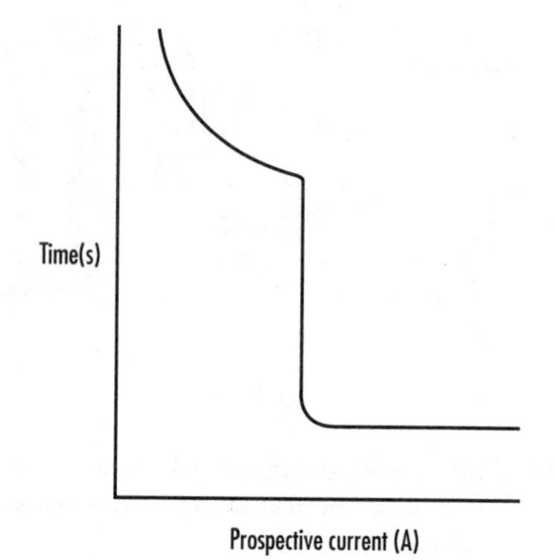

Fig. 6

25. Where equipment is used outside the equipotential zone of a building and it is supplied from a socket outlet, protection is required by a

a BS3036 rewirable fuse
b BS1361 cartridge fuse
c BS3871 miniature circuit breaker
d BS4293 residual current device.

ASSESSMENT 8.1

Answers, Hints and References

1. **a**: **b** refers to the adequacy of operation of protective devices; **c** can operate on earth faults but not short-circuit faults; and **d** creates system stability if one point of the supply transformer is earthed.
2. **b** *See* Ref. 4, definitions and Ref. 12, pp 54–59.
3. **b** *See* Ref. 4, definitions.
4. **a** *See* Ref. 4, regulation 547–02–02.
5. **c** *See* Ref. 4, Fig. 6.
6. **d**.
7. **b** *See* Ref. 12, p 62.
8. **b** *See* Ref. 4, regulation 543–02–01.
9. **b** *See* Ref. 4, definitions.
10. **c** *See* Ref. 4, definitions.
11. **a** *See* Ref. 4, definitions.
12. **d** *See* Ref. 4, regulation 547–02–02.
13. **d**.
14. **c** *See* Ref. 4, regulation 471–13–04.
15. **b** *See* Ref. 12, p 59, Fig. 4.3.
16. **c** *See* Ref. 4, regulation 471–08–06.
17. **c** *See* Ref. 4, regulation 543–02–01.
18. **c** *See* Ref. 4, regulation 514–13–01.
19. **c** *See* Ref. 4, Table 54G.
20. **a**.
21. **c** *See* Ref. 1, p 22, Fig. 2.6.
22. **d** *See* Ref. 12, p 65.
23. **a** *See* Ref. 12, p 67.
24. **b** *See* Ref. 12, pp 67–69, Fig. 4.22.
25. **d** *See* Ref. 4, group regulations 471–16.

ASSESSMENT 8.2

1. The type of earthing system shown in Fig. 1 is called a

a TN-S system
b IT system
c TT system
d TN-C-S system.

2. The **main earthing terminal** of a consumer's electrical installation is bonded to the gas and water services to

a decrease the external impedance
b create an equipotential zone
c allow protective devices to operate
d increase earth electrode resistance.

3. With reference to Fig. 1 the conductors marked gas and water are called

a supplementary bonding conductors
b main equipotential bonding conductors
c main earthing conductors
d circuit protective conductors.

4. Which of the following is an **extraneous conductive part** that is likely to introduce a potential into an electrical installation?

a central heating system
b earth electrode
c main cable armouring
d lightning conductor.

5. The British Standard number for an earth clamp for bonding purposes is

a BS3036
b BS1361
c BS951
d BS88.

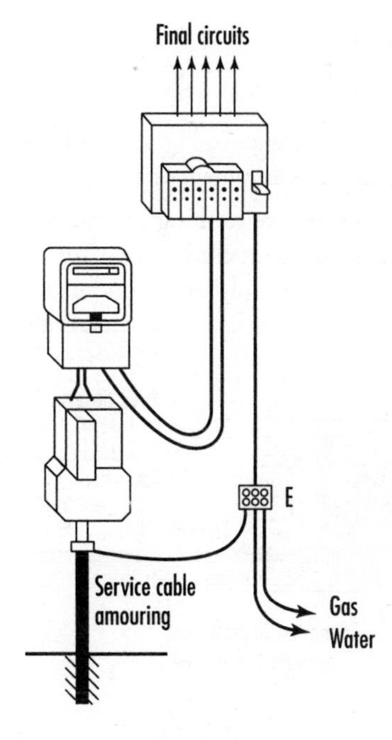

Fig. 1

6. The route taken by leakage current to earth from the point of occurrence to the supply transformer and back again is called the

a earth conductivity path
b earth fault loop impedance path
c earth return fault path
d earth let-through energy path.

7. Which one of the following earthing systems was once known as **protective multiple earthing**?

a TN-S system
b TN-C-S system
c TT system
d IT system.

8. In the **TN-C-S earthing system** shown in Fig. 2 the cut-out fuse is 100 A. What is the minimum size main equipotential bonding conductors that can be used?

a 25 mm^2
b 16 mm^2
c 10 mm^2
d 6 mm^2.

9. With reference to Fig. 2 the **PEN conductor** marked X is known as a

a protective earthing conductor
b continuous earth conductor
c mains supply earth conductor
d combined neutral/earth conductor.

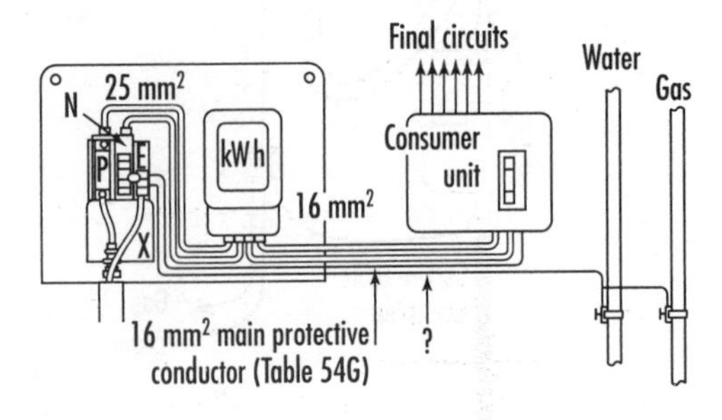

Fig. 2

10. The name of the conductor connecting together the main earthing terminal of an electrical installation to an earth electrode is called a

a supplementary bonding conductor
b main equipotential bonding conductor
c earthing conductor
d circuit protective conductor.

11. A method of protection against indirect contact, other than earthing and bonding is to provide

a barriers or enclosures
b obstacles
c electrical separation
d residual current devices.

12. Which one of the following circuit protective devices has a correction factor of 0.725?

a BS88 Part 6 cartridge fuse
b BS1361 Type 1 cartridge fuse
c BS3871 miniature circuit breaker
d BS3036 semi-enclosed rewirable fuse.

13. Where a copper earthing conductor is buried in the ground and is protected against mechanical damage but not against corrosion, its minimum size shall be

a 50 mm^2
b 35 mm^2
c 25 mm^2
d 16 mm^2.

14. Fig. 3 shows a set of time–current characteristic curves for miniature circuit breakers. The upper part of the curve is the device's

a short-circuit operation
b thermal operation
c earth-leakage operation
d open-circuit operation.

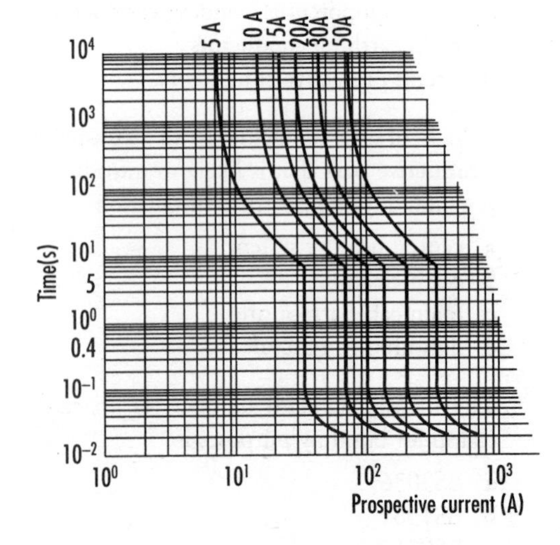

Fig. 3

15. For a **residual current device** to give supplementary protection against electric shock risk it has to operate within

a 40 ms at 30 mA
b 30 ms at 20 mA
c 20 ms at 10 mA
d 10 ms at 5 mA.

16. All of the following are instances where an installation is required to incorporate one or more residual current devices, *except*

a where the earth fault loop impedance is too high
b where the Regional Electricity Company (REC) does not supply an earthed protective conductor
c where circuits supply portable equipment outdoors
d where the system of earthing is TN-C-S.

17. Which one of the following types of circuit protective device has a rated short-circuit capacity of 16.5 kA?

a BS3036 rewirable fuse
b BS1361 Type 1 cartridge fuse
c BS88 Part 2 high breaking capacity (hbc) cartridge fuse
d BS3871 M9 miniature circuit breaker.

18. The type of miniature circuit breaker commonly chosen for a fluorescent lamp circuit is

a Type 4
b Type 3
c Type 2
d Type 1.

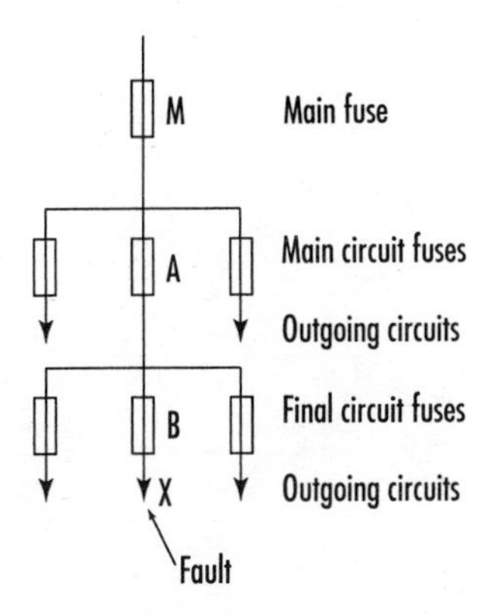

Fig. 4

19. In Fig. 4, when a fault occurs at the place marked X, discrimination of operation between fuses A and B is only possible if

a both A and B fuses rupture
b only fuse A ruptures
c only fuse B ruptures
d only fuse M ruptures.

20. In a high breaking capacity cartridge fuse, the element is surrounded by silica or graded quartz. This is done to

a strengthen the body of the fuse link
b lengthen the time for the fuse to operate
c provide an indication when the fuse ruptures
d prevent the formation of an internal arc.

21. The purpose of providing **main equipotential** and **supplementary bonding** is that

a the earth system becomes a very low impedance
b all touchable conductive parts are at earth potential
c protective devices can function satisfactorily
d water and gas pipes cannot conduct electricity.

22. All the following final circuits require a disconnection time of 0.4 s *except*

a fixed electrical equipment in bathroom installations
b fixed electrical equipment outside the equipotential zone
c portable electrical equipment which is carried while in use
d hand-held double insulated electrical equipment within the equipotential zone.

23. Fig. 5 shows a person using a hand-held portable lamp inside a restrictive conductor location. The appropriate measure of protection is by using

a separate extra low voltage
b insulated protective clothing
c a 300 mA RCD socket outlet
d Class I electrical equipment.

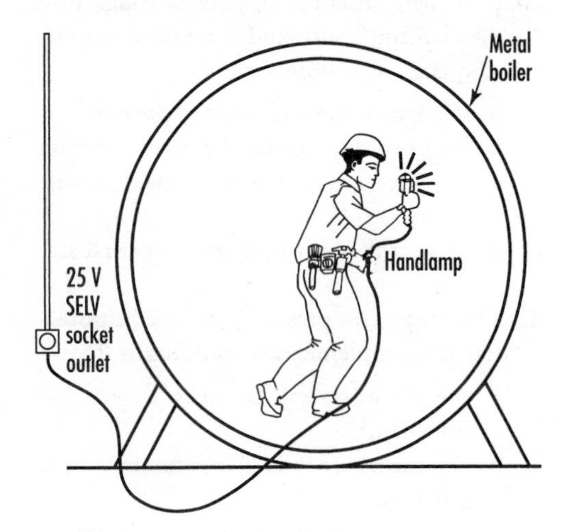

Fig. 5

24. For portable equipment used outdoors, it is a regulatory requirement for the equipment to be protected by a **residual current device** set to trip out at

a 500 mA
b 300 mA
c 50 mA
d 30 mA.

25. In a household kitchen, disconnection time of the protective device controlling the lighting circuit is limited to

a 5.0 s
b 4.0 s
c 0.4 s
d 0.2 s.

ASSESSMENT 8.2

Answers, Hints and References

1. **a** *See* Ref. 1, p 41, Fig. 3.5(a).
2. **b** *See* Ref. 4, definitions.
3. **b** *See* Ref. 4, group regulations 574–02.
4. **a** It is an independent system.
5. **c** *See* Ref. 1, p 41, Fig. 3.5(b).
6. **b** The fault current flows via the resistances of the phase and PE conductors and the reactance of the supply transformer winding, thus $Z = \sqrt{(R^2 + X^2)}$.
7. **b** *See* Ref. 4, Fig. 5, footnote.
8. **c** *See* Ref. 4, Table 54H.
9. **d** *See* Ref. 4, Fig. 5, footnote.
10. **c** *See* Ref. 4, definitions.
11. **c** *See* Ref. 4, group regulation 413–06.
12. **d** *See* Ref. 12, p 65.
13. **c** *See* Ref. 4, Table 54A.
14. **b** *See* Ref. 12, p 67.
15. **a** *See* Ref. 4, regulation 412–06–02.
16. **d** *See* Ref. 18, pp 9–10.
17. **b** *See* Ref. 18, Table 7.2A.
18. **c** *See* Ref. 18, Table 7.2B.
19. **c** Fuse B is the minor fuse nearest the fault.
20. **d** *See* Ref. 12, p 67.
21. **b** All metalwork of utility services are tied down at earth potential and should not give rise to electric shock risk by touching any exposed conductive parts.
22. **d** *See* Ref. 18, p 9.
23. **a** *See* Ref. 4, section 606.
24. **d** *See* Ref. 4, regulation 471–16–01.
25. **a** *See* Ref. 4, regulation 413–02–13.

ASSESSMENT 8.3

1. Which one of the following diagrams in Fig. 1 represents a **TN-S earthing system**?

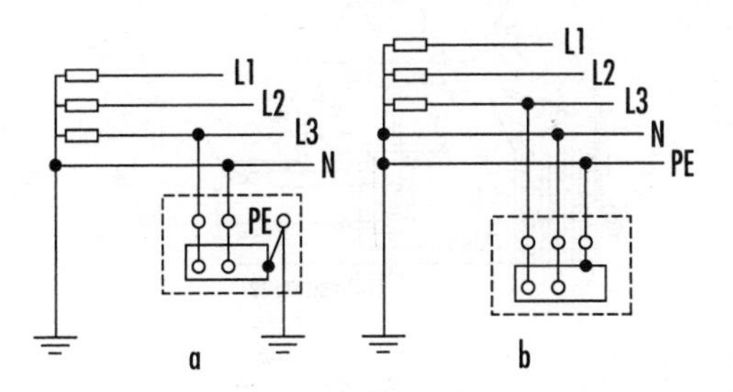

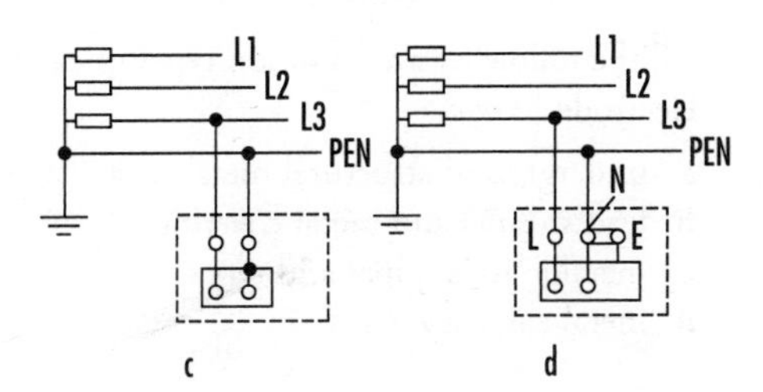

Fig. 1

2. The estimated maximum external earth impedance for a TN-C-S earthing system is

a 2.00 Ω
b 1.45 Ω
c 0.80 Ω
d 0.35 Ω.

3. The reason why items of metalwork are bonded together is to

a avoid the flow of neutral current
b ensure a common potential exists
c safeguard against corrosion
d reduce fault currents to a minimum.

4. The difference between a main **equipotential bonding** conductor and a **supplementary bonding** conductor is that the former connects together

a all metalwork of utility services to a consumer's earthing terminal
b all extraneous conductive parts to the earthing terminal
c gas and water pipes to the consumer's earth electrode
d the consumer's circuit protective conductor to the earthing terminal.

5. Earthed **equipotential bonding** and automatic disconnection of the supply is the most widely used measure of protection against

a direct contact of live parts
b indirect contact of live parts
c conductor short-circuit faults
d conductor open-circuit faults.

6. Why is it not acceptable to use a main water pipe as an earth electrode?

a the water pipe may contain insulated joints
b the local REC may refuse permission
c water is not a good conductor of electricity
d water pipes are not of a standard size.

7. A protective conductor is used to make all of the following connections *except* to

a exposed conductive parts
b Class II equipment
c earth electrodes
d extraneous conductive parts.

8. A protective conductor used in a single-core cable is made of copper for all sizes less than

a 25 mm^2
b 16 mm^2
c 10 mm^2
d 6 mm^2.

9. The colours green and yellow identify a protective conductor. What is the *least* percentage covering of any one of these colours over the cable's surface area?

a 50%
b 45%
c 30%
d 25%.

10. The star point of a supply transformer is earthed in order to make the supply network safe and also to

a stabilize the voltage
b discharge lightning strikes
c obtain a neutral return
d obtain red, yellow and blue phases.

11. If no mechanical protection is provided a supplementary bonding conductor connecting two exposed conductive parts together has to be of a size *not* less than

a 10 mm^2
b 6 mm^2
c 4 mm^2
d 2.5 mm^2.

12. Which one of the following letters of the alphabet is used as a material factor symbol?

a *h*
b *j*
c *k*
d *M*.

13. Which of the following earthing systems is mostly likely to be chosen for an electrical installation supplied by a two-wire overhead supply?

a TN-S system
b TT system
c TN-C-S system
d TN-C system.

14. In a TT **earthing system**, all socket outlets must be protected by a

a residual current device
b miniature circuit breaker
c lightning discharge arrestor
d BS3535 isolating transformer.

15. Fig. 2 shows the inside working parts of a miniature circuit breaker. What is the function of the electromagnet?

a detects open circuits
b detects short circuits
c detects slow overloads
d detects loss of voltage.

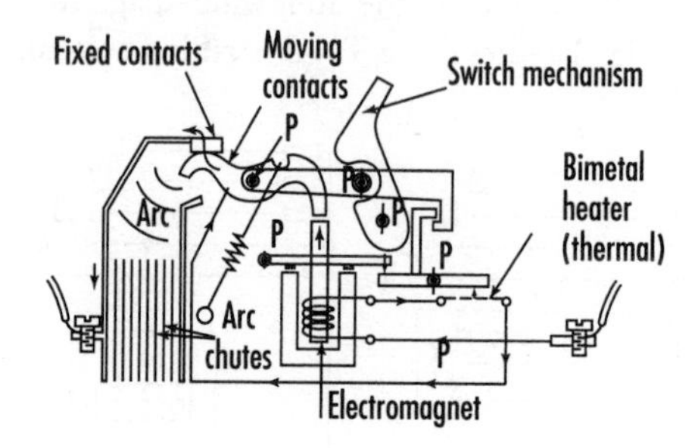

Fig. 2

16. All the following are common types of earth electrode, *except*

a underground structural metalwork
b lead sheaths and metal coverings
c metallic rods, pipes and tapes
d metal catenary wires.

17. The minimum cross-sectional area of the main equipotential bonding conductor in relation to a 35 mm^2 supply neutral is

a 25 mm^2
b 16 mm^2
c 10 mm^2
d 6 mm^2.

18. Fig. 3 shows different characteristics of miniature circuit breakers. What would be the end of range tripping value of a 15 A Type B device?

a 100 A
b 75 A
c 45 A
d 15 A.

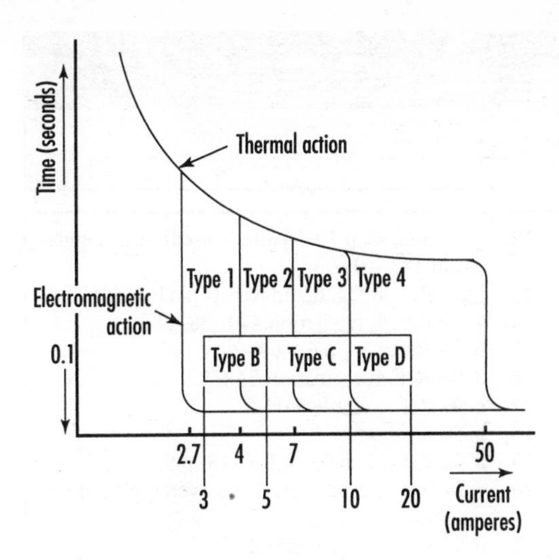

Fig. 3

19. The term **fusing factor** refers to the

   a ratio of the minimum fusing current and the current rating
   b rupturing or breaking capacity of a semi-enclosed fuse
   c correction factor of 0.725 given to a semi-enclosed fuse
   d comparative tripping current of a miniature circuit breaker.

20. In Fig. 4, to obtain ideal **fuse discrimination**

   a the main fuse should be a semi-enclosed type and all others should be cartridge fuses
   b only fuses of different make and rating should be in series with each other
   c all fuses should be made by one manufacture and time/current curves should not overlap
   d semi-enclosed fuses should not be used in series with other types of fuse.

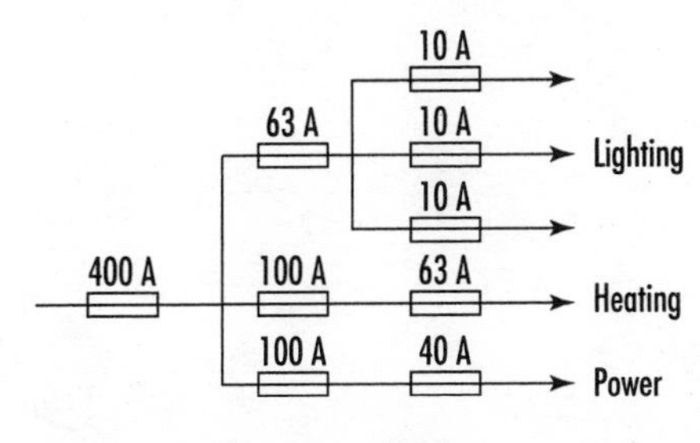

Fig. 4

21. Calculate the size of protective conductor (S) using the formula $S = \sqrt{(I^2t)}/k$, if I = 400 A, t = 0.35 s and k = 115.

   a 1.00 mm^2
   b 0.82 mm^2
   c 0.75 mm^2
   d 0.65 mm^2.

22. One of the requirements for a BS3535 shaver supply unit is that it must have

   a an earth connection made to the circuit protective conductor
   b no earth connection attached to its metal frame
   c both the primary and secondary windings earthed
   d its secondary winding earthed at the mid-point only.

23. Fig. 5 shows an earth electrode's concrete inspection pit. What requirement is missing?

   a separate test terminal
   b contractor's nameplate
   c earth connection label
   d in-fill of sand.

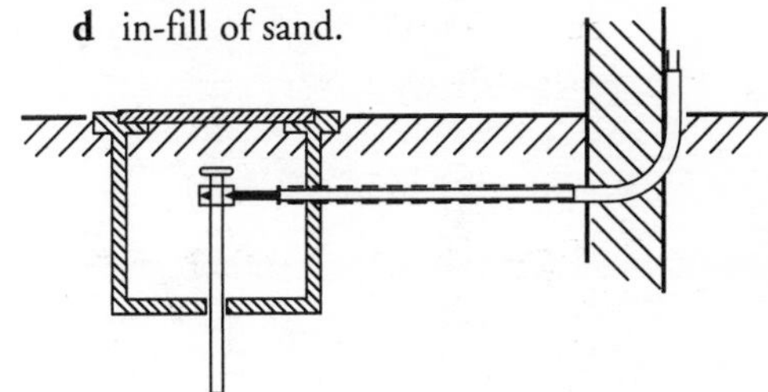

Fig. 5

24. Which one of the following classes of equipment requires to be earthed?

   a Class 0 equipment
   b Class I equipment
   c Class II equipment
   d Class III equipment.

25. Where a 240 V mains connected high-pressure water hose is used outside a building, protection is required by a residual current device having a tripping current not less than

   a 500 mA
   b 300 mA
   c 100 mA
   d 30 mA.

ASSESSMENT 8.3

Answers, Hints and References

1. **b** *See* Ref. 4, definitions.
2. **d** *See* Ref. 18, p 1.
3. **b** Often referred to as earth potential. By doing this the impedance of the system will be low and any fault to earth will encourage a large current to flow and disconnect circuit protective devices.
4. **a** *See* Ref. 4, Fig. 2.
5. **b** *See* Ref. 4, group regulations 413–02.
6. **a** *See* Ref. 4, definitions.
7. **b** *See* Ref. 4, definitions.
8. **c** *See* Ref. 4, regulation 543–02–01.
9. **c** *See* Ref. 4, regulation 514–03–01.
10. **a** The line and phase voltages are tied down to a common value.
11. **c** *See* Ref. 4, regulation 547–03–01.
12. **c** *See* Ref. 4, p 18 (symbols used) and, regulation 543–01–03.
13. **b** *See* Ref. 4, Fig. 6 and Ref. 1, p 41, Fig. 3.5(a).
14. **a** *See* Ref. 4, regulation 471–08–06.
15. **b** *See* Ref. 12, pp 67–69.
16. **d** *See* Ref. 4, regulation 542–02–01.
17. **c** *See* Ref. 4, Table 54H.
18. **b** $15\ \text{A} \times 5 = 75\ \text{A}$.
19. **a** *See* Ref. 12, p 65 or Ref. 18, p 96.
20. **c** *See* Ref. 4, Appendix 3 (no overlapping time-current curves).
21. **d** $S = (\sqrt{(400^2 \times 0.035)}/115)$
22. **b** *See* Ref. 4, regulation 601–09–01.
23. **d** *See* Ref. 4, regulation 514–13–01.
24. **d** *See* Ref. 4, definitions.
25. **d** *See* Ref. 4, group regulations 471–16.

9

PROCEDURES AND PRACTICES FOR THE INSPECTION AND TESTING OF ELECTRICAL INSTALLATION

To tackle the assessments in Topic 9 you will need to know:

- the procedure for isolating parts of a consumer's electrical wiring prior to making certain inspections and tests;
- the various types of inspection and test forms that need to be made available to relevant personnel;
- the relevant statutory and non-statutory requirements for carrying out inspections and making tests;
- the inspection and testing requirements for domestic, commercial and industrial premises;
- the location of where to make inspections and tests from information provided by drawings, specifications, etc;
- the test equipment to be used and the sequence for making inspections and tests;
- the procedure for recording test results and reporting unsatisfactory test results.

DEFINITIONS

Circuit chart – a list of important items of information such as number of points, size and type of cable, circuit protection, type of wiring, etc.
Continuity test – a test made to check the 'soundness' of a circuit conductor, often using an ohmmeter which measures the conductor's resistance value.
Earth leakage current – current escaping to the general mass of earth often because of insulation breakdown.
Earth fault loop impedance test – a live test on final circuits to see if they have been properly designed so that circuit protective devices will disconnect in appropriate times.
Insulation resistance test – a high ohmic resistance test carried out to see if conductors, circuits and equipment have sound insulation.

Insulation test

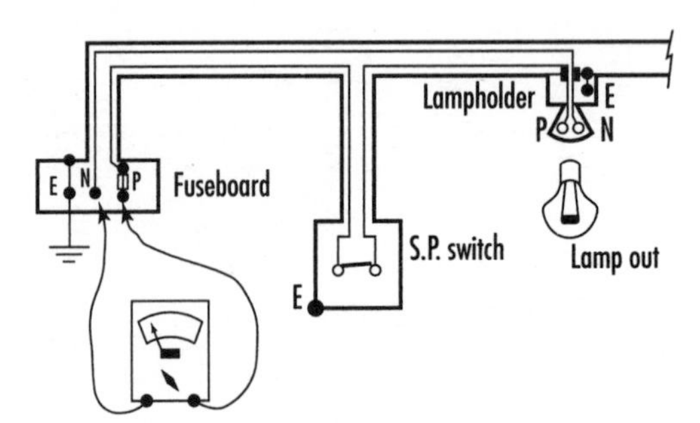

Multiple-loop connection – a wiring error in a 30 A ring final circuit whereby one or more socket-outlets have been accidentally shorted.
Ohmmeter – a measuring instrument for finding high and low values of resistance.
Open circuit – a condition whereby a conductor or circuit will not allow current to flow, either as a result of a fault or deliberate action through some form of switch or circuit breaker.
Periodic inspection – a procedure made by a test inspector for reporting the condition of an existing installation. An example might be to check measures of direct and indirect contact or circuit protective devices switchgear labelling, diagrams, schedules, etc.
Periodic testing – a procedure made by a test inspector for reporting the condition of an existing installation. An example might be to make continuity, polarity, insulation resistance, earth fault loop impedance and other types of test.
Phase conductor – one of three conductors derived from a three-phase supply system that when run with a neutral conductor provides a single-phase supply of 240 V.
Polarity test – a test on a 'live' or 'dead' circuit to find the correct polarity of the conductors.
Potentially dangerous atmosphere – an atmosphere that contains a vapour or gas likely to give rise to an explosion.
Proving dead – a procedure to determine whether a circuit or equipment is 'live' after it has already been switched off or been isolated.

Three-phase working

Residual current device – a current-operated earth leakage circuit breaker designed to trip out when it senses an earth fault.
Ring circuit continuity test – a test method to determine the ohmic value of all the circuit conductors in a ring final circuit and to see if they are continuous.

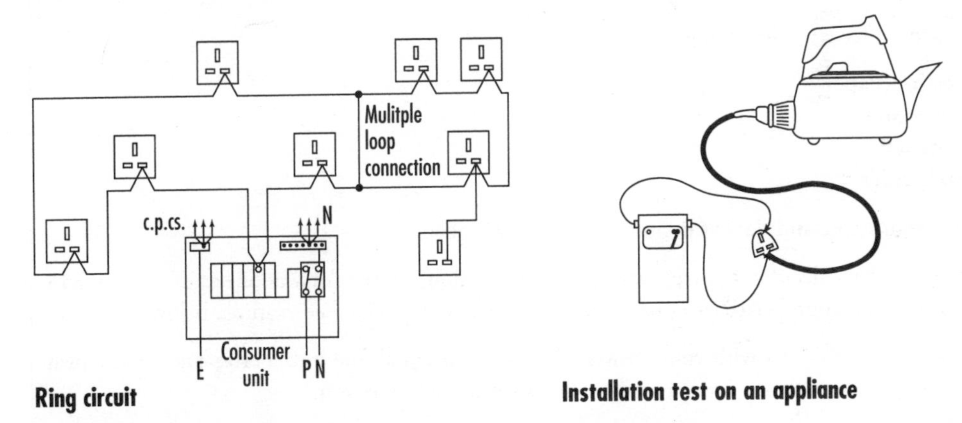

Ring circuit

Installation test on an appliance

Statutory regulation – a mandatory regulation passed by an act of Parliament.
Test link – a connection block to make tests on an earth electrode without interfering with the consumer's earthing terminal.
Tong tester – a clamp-on ammeter measuring instrument.
Visual inspection – a procedure to examine samples of electrical wiring to verify that they have been selected and erected correctly and comply with applicable British Standards and other standards. The test is also done to look for damage and that the wiring and equipment is suitable for the environmental conditions.

FORM WR5 Form No xxxx

PERIODIC INSPECTION REPORT FOR AN ELECTRICAL INSTALLATION (BS 7671:1992) Note 1

Details of the Client

Client

Address

Purpose for which this Report is required

DETAILS OF THE INSTALLATION

Occupier:

Address

Description of Premises: Domestic Commercial Industrial Other:

FORM WR3 Form No xxxx

TEST SCHEDULE

SCHEDULE OF ITEMS TO BE TESTED

1. ☐ Continuity of protective conductors
2. ☐ Continuity of ring final circuit conductors
3. ☐ Insulation resistance between live conductors and earth
4. ☐ Site applied insulation
5. ☐ Protection by separation of circuits
6. ☐ Protection against direct contact, by barrier or enclosure provided during erection
7. ☐ Insulation of non-conducting floors and walls
8. ☐ Polarity
9. ☐ Earth electrode resistance
10. ☐ Earth fault loop resistance
11. ☐ Operation of residual current-operated devices
12. ☐ Functional testing of assemblies

Tick items meeting test requirements. Delete tests not applicable.

DEPARTURES FROM REGULATIONS (note should be made of compliance with relevant tests where appropriate).

Inspection and test forms

TOPIC 9

Procedures and Practices for the Inspection and Testing of Electrical Installation

ASSESSMENTS 9.1 – 9.3

Time allowed: 1 hour

Instructions

* You should have the following:

  Question Paper
  Answer Sheet
  HB pencil
  Metric ruler

* Enter your name and date at the top of the Answer Sheet.
* When you have decided a correct response to a question, on the Answer Sheet, draw a straight line across the appropriate letter using your HB pencil and ruler (see example below).
* If you make a mistake with your answer, change the original line into a cross and then repeat the previous instruction. There is only one answer to each question.
* Do not write on any page of the Question Paper.
* Make sure you read each question carefully and try to answer all the questions in the allotted time.

Example:

a 400 V	**a** 400 V
~~**b**~~ 315 V	~~**b**~~ (crossed) 315 V
c 230 V	**c** 230 V
d 110 V	~~**d**~~ 110 V

ASSESSMENT 9.1

1. Which one of the following regulations is non-statutory?

a Factory Act Special Regulations
b Electricity at Work Regulations
c IEE Wiring Regulations
d Electricity Supply Regulations.

2. To avoid danger, it is *not* safe to make a periodic inspection and test on an electrical installation until all relevant parts have been properly

a isolated from the electricity supply
b reported to the owner of the premises
c sectioned off from the premises owner
d earthed and bonded at the intake point.

3. Before work commences on an electrical circuit that is to be isolated, the term **proving dead** means that the circuit

a will be tested to see if it is still alive
b switch has been observed in the off position
c has been labelled with a warning notice
d cannot be interfered with by other persons.

4. All of the following information is required by a test inspector carrying out inspection and testing work of an electrical installation *except* the

a type of earthing arrangements
b installation user's safety policy
c number and type of live conductors
d maximum demand stated in amperes.

5. The procedure for safe isolation of an electrical installation, before making non-energized tests, should include all of the following steps, *except*

a 'locking off' of switches
b selecting correct equipment
c provision of warning notices
d taking verbal instructions.

6. In Fig. 1, the electrician is working on the centre phase conductor of a 400 V three-phase ac supply. In the normal sequence, the phase colour of this live conductor is

a red
b yellow
c white
d blue.

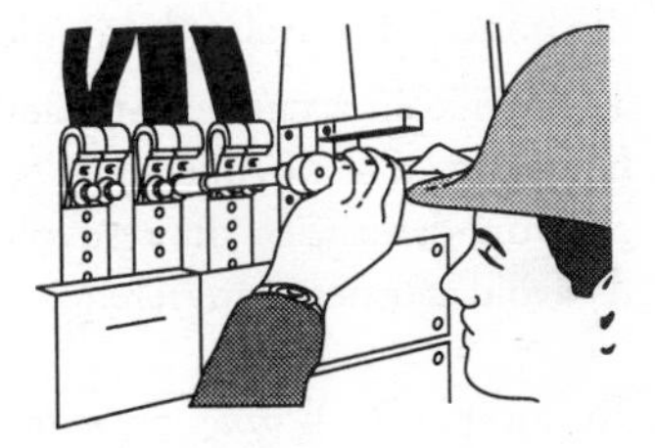

Fig. 1

7. An electrical installation should be arranged so as to avoid danger in the event of a fault and also to allow safe operation, inspection and maintenance. One method of complying with this is to

a divide the system into separate circuits
b connect all final circuits as ring mains
c install several residual current devices
d design all final circuits with miniature circuit breakers.

8. Which one of the following is *not* a reason for making an inspection of an electrical installation?

a checking for signs of any visible damage that might impair safety
b checking to see if equipment has been properly selected and erected
c checking to see if equipment complies with recognized standards
d checking for information concerning the reliability of protective devices.

9. When inspecting final circuits in domestic premises it would be unusual for a test inspector to come across a

a BS3871 Type 4 miniature circuit breaker
b BS3036 rewirable fuse
c BS1361 cartridge fuse
d BS1362 cartridge fuse.

10. In which installation environment listed below are you most likely to find a PVC armoured cable?

a general lighting in an office
b power supply to a factory
c submerged garden pool lights
d switchgear interconnections.

11. When making an insulation resistance test, it is important for electronic devices to be disconnected or isolated, in order to

a provide an accurate test reading
b provide a test frequency at 50 Hz
c avoid damage by the test voltage
d avoid transient interference.

12. To meet *IEE Wiring Regulations* requirements, which one of the following set of test numbers indicates the correct sequence of tests to be made?

(1) site-applied insulation
(2) insulation resistance
(3) continuity of protective conductor
(4) continuity of ring final circuit conductors.

a 4, 3, 2, 1
b 3, 4, 2, 1
c 2, 3, 4, 1
d 1, 2, 3, 4.

13. Fig. 2 shows two methods of making an insulation resistance test on a domestic consumer's installation. In Test 2

a the ohmmeter reading must not be less than 5 MΩ
b all electrical equipment must be switched off or disconnected
c all protective devices must be switched off
d the instrument must generate an ac voltage of 500 V.

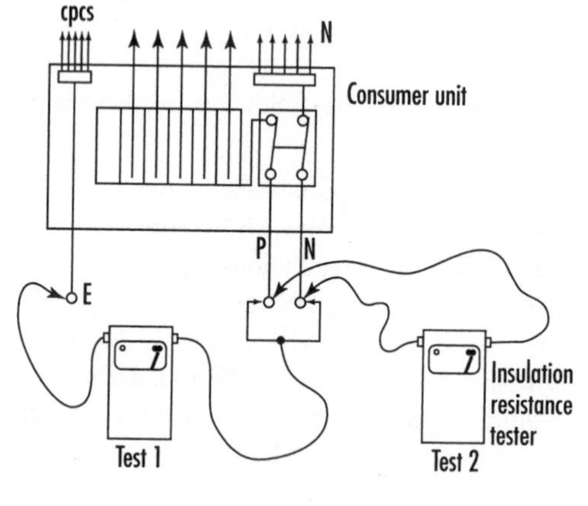

Fig. 2

14. In Fig. 3, one reason for making a test of ring final circuit continuity is to see if there

a is no wrong polarity connection
b is a resistance above 0.5 Ω
c are any likely disconnections
d are any branch cables connected.

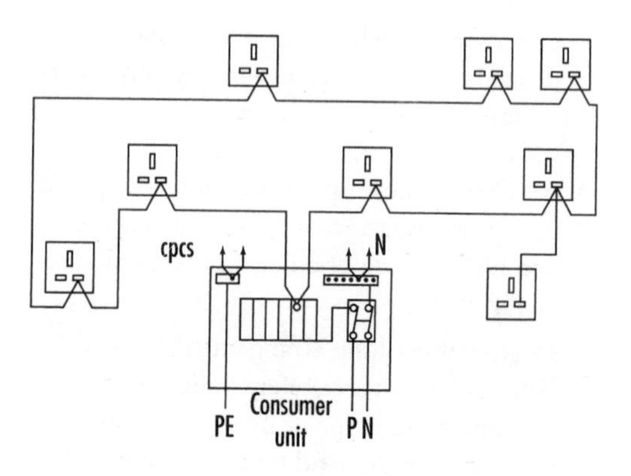

Fig. 3

15. Which one of the electrical installations listed below suffers most from problems caused by dampness and corrosion?

a milking parlour
b boiler house
c domestic dwelling
d office block.

16. Any outside changes that are likely to affect an electrical installation's design and safe operation are called

a environmental factors
b exposure limits
c detrimental fluctuations
d external influences.

17. Which one of the following requirements does *not* fall within the detailed inspection of installed electrical equipment prior to the testing of an installation?

a no signs of visible damage or defect
b presence of a recognized mark or certificate
c satisfactory selection and erection
d product manufacturer is Part 1 BS5750.

18. It is necessary to liaise with occupants of a premises prior to commencing periodic inspection and testing. This is to

a allow movement of people to other areas of work
b create a good relationship with employer and employees
c avoid supplies being switched off without notice
d solve problems with managers and trade union officials.

19. All of the following have to be verified for correct polarity, *except*

a connections inside socket outlets
b connections inside Edison-screw lampholders
c connections inside single-pole switches
d connections inside bayonet-cap lampholders.

20. One of the items that should be included in an installation schedule is

a details of the client's contract
b the location of all portable equipment
c the number of outlets in the installation
d all local switching arrangements.

21. It is recommended that domestic premises be inspected and tested at intervals no longer than

a 10 years
b 5 years
c 3 years
d 2 years.

22. When completing an electrical installation certificate, all of the following areas need to be signed by a competent person, *except*

a electrical design
b contract conditions
c construction
d inspection and test.

23. Which one of the following is *not* a relevant form required by a test inspector prior to making inspection and tests of an electrical installation?

a a completion and inspection certificate
b test results and installation schedule
c particulars of the installation
d standard building contract conditions.

24. Which one of the following tests does *not* need the recording of ohmic results?

a polarity of conductors
b insulation resistance
c ring circuit continuity
d earth fault loop impedance.

25. All of the following are common deviations from the *IEE Wiring Regulations*, *except*

a omission of a quarterly rcd notice
b equipment not suitable for ac voltages
c ingress of water in enclosures
d presence of a 415 V notice.

ASSESSMENT 9.1

Answers, Hints and References

1. **c** *See* Ref. 4, Appendix 2.
2. **a** *See* Ref. 2, pp 54–55.
3. **a** *See* Ref. 20, pp 9–11.
4. **b** *See* Ref. 28, no.3, p 3.
5. **d** *See* Ref. 20, pp 9–11.
6. **b** *See* Ref. 4, Table 51A.
7. **a** *See* Ref. 4, section 314.
8. **d** It is concerned with product quality.
9. **a** *See* Ref. 12, p 67.
10. **b**.
11. **c** *See* Ref. 1, p 86.
12. **b** *See* Ref. 4, Chapter 71.
13. **b** *See* Ref. 1, pp 85–86.
14. **c** *See* Ref. 1, p 82, Fig. 4.3.
15. **a**.
16. **d** *See* Ref. 4, definitions.
17. **d** BS5750 refers to Quality Assurance.
18. **c**.
19. **d**.
20. **c** *See* Ref. 4, Section 314.
21. **a** *See* Ref. 2, p 5, Table 1.1.
22. **b** *See* Ref. 18, p 100.
23. **d**.
24. **a**.
25. **d** *See* Ref. 4, group regulations 514–10.

ASSESSMENT 9.2

1. Which one of the following regulations is *specifically relevant* to inspection and testing of premises?
   - **a** Electricity Supply Regulations
   - **b** Electricity at Work Regulations
   - **c** IEE Wiring Regulations
   - **d** Factory Act Special Regulations.

2. Checking equipment for damage, to ensure that it has been properly installed and that it complies with relevant standards, is the main reason for
   - **a** a one-year warranty
   - **b** verifying its CE mark
   - **c** testing at 500 V dc
   - **d** initial inspection.

3. All of the following are methods of isolation, *except*
   - **a** switch disconnector
   - **b** limit switches
   - **c** fuse links
   - **d** plugs and socket-outlets.

4. The term that describes a live test procedure before work commences on an electrical circuit is known as
   - **a** proving dead
   - **b** supply identification
   - **c** polarity assessment
   - **d** double checking.

5. All of the following are relevant forms required by an inspector prior to carrying out inspection and testing of a premises, *except* the form dealing with
   - **a** completion and inspection certificate
   - **b** test results and installation schedule
   - **c** particulars of the installation
   - **d** standard building contract conditions.

6. Where would a test inspector come across a BS3871 Type 3 miniature circuit breaker being used?
   - **a** motor circuit
   - **b** immersion heater circuit
   - **c** tungsten lamp circuit
   - **d** welding circuit.

7. A test inspector is most likely to find PVC sheathed MI cable installed in a
   - **a** damp environment
   - **b** dry environment
   - **c** submerged environment
   - **d** hot environment.

8. In which premises are fire barriers likely to be checked?
   - **a** open substation buildings
   - **b** cathedrals over 50 m high
   - **c** farm yard installations
   - **d** multistorey office blocks.

9. When making a **ring circuit continuity** test, it has to be carried out on

a only the phase and circuit protective conductors
b only the phase and neutral conductors
c the phase, neutral and circuit protective conductors
d the main tails connecting the consumer unit.

10. In terms of electrical ignition from a fault, all of the following premises could have potentially dangerous atmospheres, *except* a

a refrigeration plant
b works' paint shop
c chemical works
d garage repair workshop.

11. What is the name of the test that is used for finding out the presence of **earth leakage current**?

a continuity resistance test
b conductor polarity test
c insulation resistance test
d earth loop impedance test.

12. An insulation resistance test is carried out on a completed 240 V installation. Which one of the following would be the minimum acceptable value?

a 2.0 MΩ
b 1.0 MΩ
c 0.5 MΩ
d 0.2 MΩ.

13. In Fig. 1 the item marked X is called a **test link** and has to be disconnected when the earth electrode is being tested. This is to make sure that the test

a cannot affect other installations
b results are not invalidated
c instrument can be freely connected
d cannot harm the test inspector.

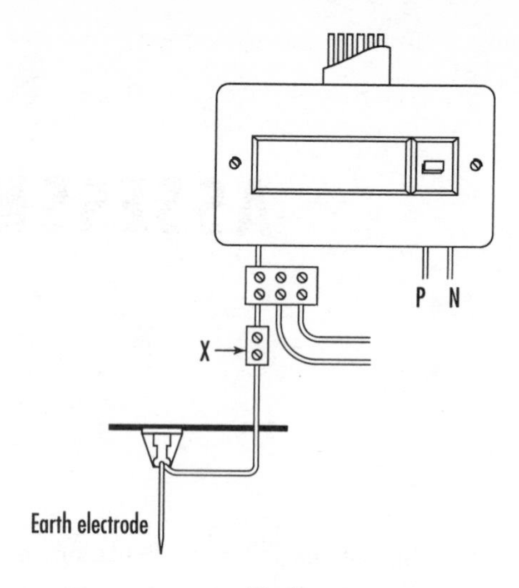

Fig. 1

14. The ohmmeter instrument used for making an insulation resistance test on a 415 V installation should be capable of generating a dc test voltage of

a 1000 V
b 650 V
c 500 V
d 250 V.

15. What type of test is being carried out in Fig. 2?

a impedance test
b voltage test
c continuity test
d earth test.

16. In Fig. 2 the ohmmeter should be capable of passing a test current greater than

a 200 kA
b 200 A
c 200 mA
d 200 μA.

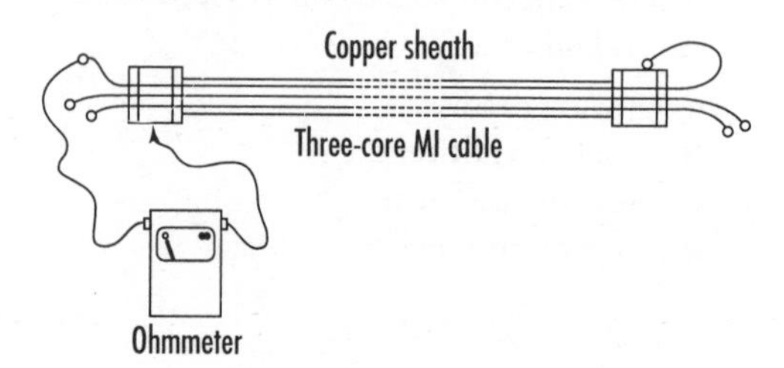

Fig. 2

17. Which one of the following is *not* a verification of polarity?

- **a** correct connections inside a 13 A socket outlets
- **b** correct live and neutral connections inside Edison-screw lampholders
- **c** phase conductors connections inside single-pole switches
- **d** correct order of circuit conductors inside a fuseboard.

18. In Fig. 3 if terminals 2 and 3 are joined together, on what other terminals would you connect a low reading ohmmeter to find the continuity between live conductors?

- **a** 1 and 2
- **b** 3 and 4
- **c** 4 and 5
- **d** 5 and 6.

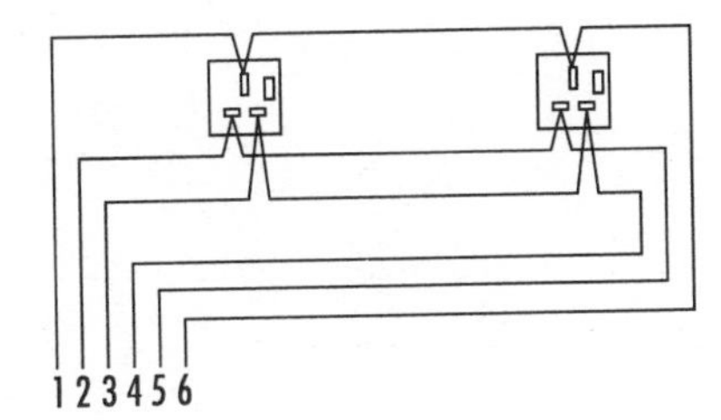

Fig. 3

19. Which one of the following tests requires ohmic results in the order of millions of ohms?

- **a** polarity of conductors
- **b** insulation resistance
- **c** ring circuit continuity
- **d** earth fault loop impedance.

20. If the element of an electric fire is found to be live after its switch has been placed in the off position, the fire is likely to have

- **a** a leakage to earth fault
- **b** a faulty element connection
- **c** no earthed metal connection
- **d** reverse conductor polarity.

21. Fig. 4 shows an Edison-screw lampholder. The centre terminal should be connected to the

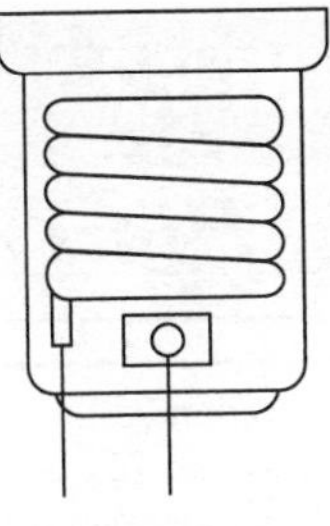

Fig. 4

- **a** live circuit conductor labelled red
- **b** circuit conductor labelled black
- **c** switch-feed conductor labelled red
- **d** switch-wire conductor labelled red.

22. The purpose of a **circuit chart** found inside a distribution fuseboard is to provide information about

- **a** circuit routes, types of circuit protection and methods of wiring
- **b** types of circuit protective device that can be used alternatively
- **c** parts of each circuit that have been inspected for polarity
- **d** circuit deviations and remedial procedures that have been noted.

23. A **polarity test** on an installation is to find out if

- **a** electric motors run the correct way
- **b** electrical equipment is suitable for ac voltage
- **c** the wiring circuits function correctly
- **d** switches are connected in the phase conductors.

24. The *IEE Wiring Regulations* require a residual current device to be tested

- **a** weekly
- **b** monthly
- **c** quarterly
- **d** yearly.

25. It is recommended that an agricultural premises be periodically inspected and tested every

- **a** 10 years
- **b** 5 years
- **c** 3 years
- **d** 2 years.

ASSESSMENT 9.2

Answers, Hints and References

1. **c** *See* Ref. 4, section 7.
2. **d** *See* Ref. 4, section 712.
3. **b** The switch contact gap is too small.
4. **a** *See* Ref. 20, pp 9–11.
5. **d** *See* Ref. 28, no.3.
6. **a** *See* Ref. 12, p 67.
7. **a** *See* Ref. 18, Table 3A.
8. **d** *See* Ref. 12, p 76, Fig. 5.3.
9. **c** *See* Ref. 18, p 52.
10. **a** *See* Ref. 2, p 62.
11. **c** *See* Ref. 1, p 85.
12. **c** *See* Ref. 4, Table 71A.
13. **b** *See* Ref. 18, pp 60–61.
14. **c** *See* Ref. 4, regulation 713–08–02.
15. **c** *See* Ref. 1, p 80, Fig. 4.2.
16. **c** *See* Ref. 4, regulation 713–02–01.
17. **d** *See* Ref. 1, p 86.
18. **c**.
19. **b**.
20. **d**.
21. **d** *See* Ref. 1, p 86.
22. **a** *See* Ref. 4, regulation 514–09–01.
23. **d** *See* Ref. 1, p 86.
24. **c** *See* Ref. 4, regulation 514–12–02.
25. **c** *See* Ref. 2, Table 1.1, p 5.

ASSESSMENT 9.3

1. The reason why parts of an electrical installation are required to be isolated prior to carrying out inspection and testing is to

a check the loading of different final circuits
b avoid the risk of injury or electrocution
c see if electrical circuits are adequately protected
d see if adequate switching exists within the system.

2. The main purpose of inspecting and testing an electrical installation is to see if the wiring and installed equipment are both

a capable of functioning under normal and abnormal conditions
b safe and comply with various regulation requirements
c manufactured to a satisfactory standard of product quality
d safe to use for a minimum period of 20 years.

3. The correct method of checking if a circuit is safe to work on is to

a switch off the supply mains to the whole installation
b connect an appliance and test its operation
c trace the wiring and remove the fuse
d use an approved test lamp and secure isloation.

4. All of the following are forms to be used by the person carrying out inspection and testing work on an electrical installation *except*

a installation schedule
b completion certificate
c user's operating manual
d periodic inspection report.

5. Which person needs to know of the recommendations for periodic inspection and testing?

a consumer ordering the work
b inspector making the report
c inspector from the local Regional Electricity Company
d surveyor from Local Authority.

6. Which one of the following tests on an electrical installation should *not* be carried out until a supply is connected?

a insulation resistance
b ring circuit continuity
c earth fault loop impedance
d protective conductor continuity.

7. Fig. 1 shows how an insulation resistance test is made. What would be the ohmmeter's reading after closing the local switch and inserting the lamp?

a it would not be able to supply 500 V
b it would read infinity
c it would not read in megohms
d it would read a 'dead' short.

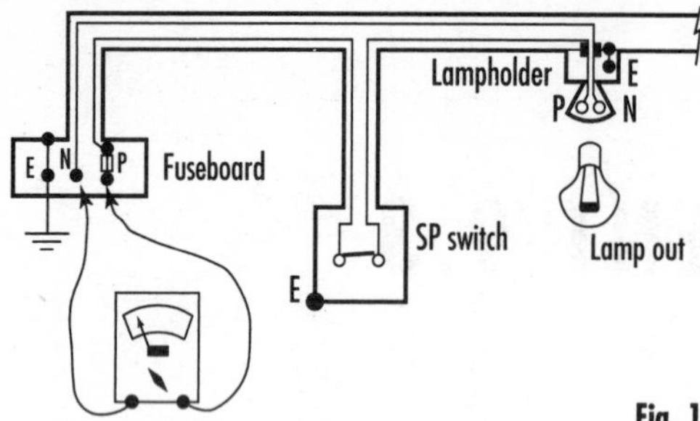

Fig. 1

8. In Fig. 1, if the lamp was inserted and the insulation test made, what would be the effect of opening the local switch?

- **a** the ohmmeter would read infinity on the megohm scale
- **b** the ohmmeter would read zero ohm
- **c** the ohmmeter would only read the lamp's resistance
- **d** the ohmmeter would read an open circuit voltage of 500 V.

9. Name the **statutory regulations** that require a maintenance record to be made of all electrical equipment.

- **a** Electricity Act
- **b** Electricity (Factory Act) Special Regulations
- **c** Electricity Supply Regulations
- **d** Electricity at Work Regulations.

10. In Fig. 2, for a complete test of **phase conductor polarity** the local switch needs

- **a** opening and a link inserted between P and PE
- **b** opening and a link inserted between P and N
- **c** closing and a link inserted between P and PE
- **d** closing and a link inserted between P and N.

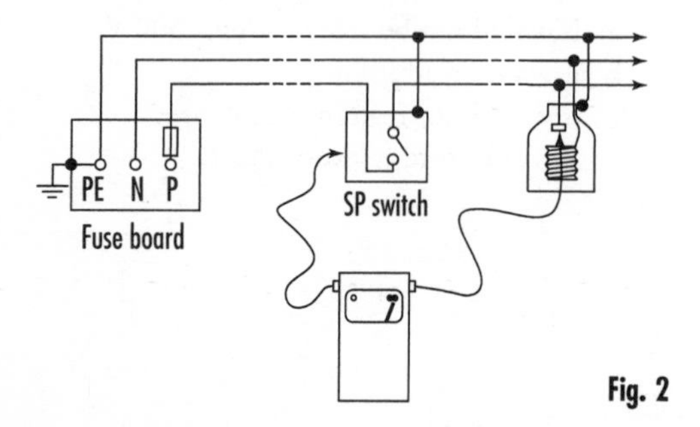

Fig. 2

11. In Fig. 2, all of the following instruments could be used for finding the polarity, *except* a

- **a** bell tester
- **b** tong tester
- **c** ohmmeter
- **d** multimeter.

12. An electrical installation should be designed so as to avoid loss of supply in the event of a fault. One method of complying with this is to

- **a** arrange all final circuits with circuit breakers
- **b** connect all final circuits as ring mains
- **c** install a number of residual current devices
- **d** divide the installation into separate circuits.

13. When carrying out **visual inspection** of an installation, which of the following needs to be verified?

- **a** prospective short circuit current
- **b** rotation of the phase conductors
- **c** presence of fire barriers
- **d** maximum permissible voltage drop.

14. On completion of an electrical installation an insulation resistance test of conductors to earth is to be made. For this test it is important to

- **a** join together all neutral and protective conductors
- **b** place all local switches in the 'off' position
- **c** insert all lamps in their ceiling roses
- **d** insert or switch on all circuit protective devices.

15. Fig. 3 shows a method of making an earth continuity test. The alphabet symbols used to abbreviate the resistance values of the phase conductor and circuit protective conductor (cpc) are, respectively

- **a** A_1 and A_2
- **b** P_R and P_E
- **c** R_1 and R_2
- **d** P_C and E_C.

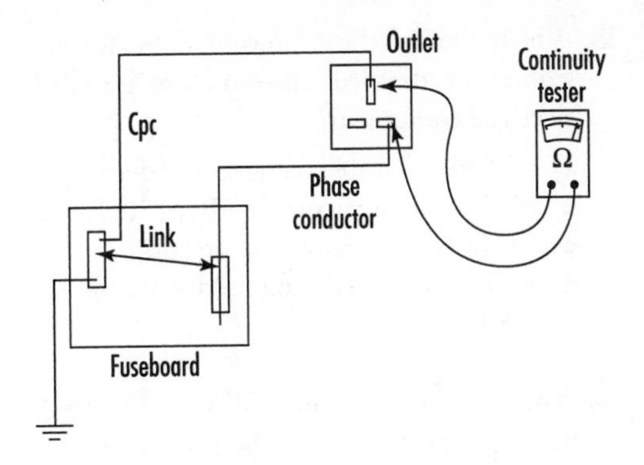

Fig. 3

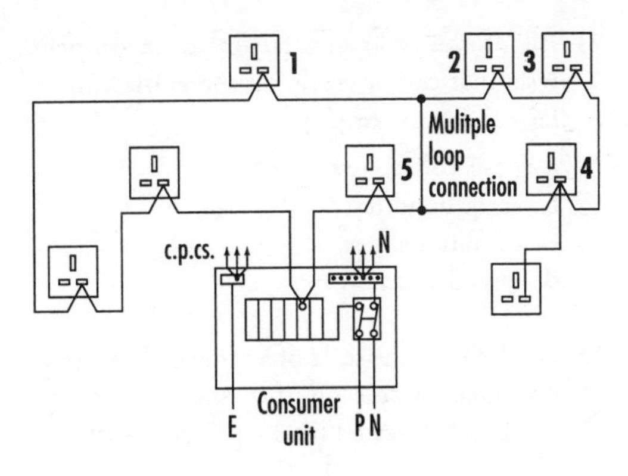

Fig. 4

16. With reference to Fig. 3, if the length of circuit was 8 m, what would be the reading on the tester (ignoring test leads) if the phase conductor and cpc had resistance values of 1.83 mΩ/m and 4.61mΩ/m, respectively?

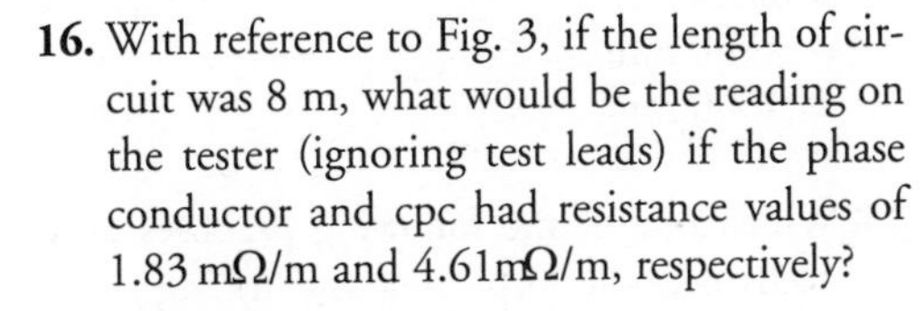

a 0.1 Ω
b 0.52 Ω
c 0.064 Ω
d 0.01 Ω.

17. If a test inspector came across a 25 mm open rubber grommet in the top of a consumer unit, it would contravene regulation requirements for barriers and enclosures by not being to the standard of

a IP2X
b IP3X
c IP4X
d IP5X.

18. Fig. 4 shows a **multiple-loop connection** in the phase conductor of a ring final circuit between socket-outlet 1 and socket-outlet 5. If the resistance of the phase conductor (and loop) between these two sockets was 1 Ω and the resistance of the phase conductors around socket-outlets 2, 3 and 4 back to the loop was 3 Ω, what would be the equivalent resistance of the multiple-loop connection?

a 4.00 Ω
b 3.50 Ω
c 0.75 Ω
d 0.33 Ω.

19. In Question 18 and Fig. 4, what is another reason for making a test of ring circuit continuity?

a detects wrong polarity
b detects open circuit
c detects closed circuit
d detects high impedance.

20. The tester shown in Fig. 5 is checking the insulation between the neutral conductor and earth. Unless otherwise stated in a British Standard, an acceptable reading to meet the requirements of the *IEE Wiring Regulations* must be above

a 5.0 MΩ
b 1.0 MΩ
c 0.5 MΩ
d 0.1 MΩ.

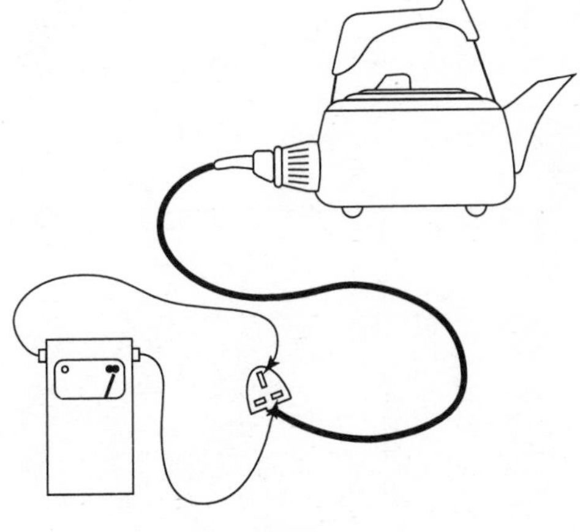

Fig. 5

21. All of the following are other instrument tests that can be made on the kettle and its lead in Fig. 5, *except*

a flash voltage test
b earth bond test
c operational test
d impedance test.

22. Fig. 6 shows the scale of a typical ohmmeter. An **open circuit** on the continuity range would be indicated by the pointer reading

a infinity
b 500 Ω
c 0.1 MΩ
d zero.

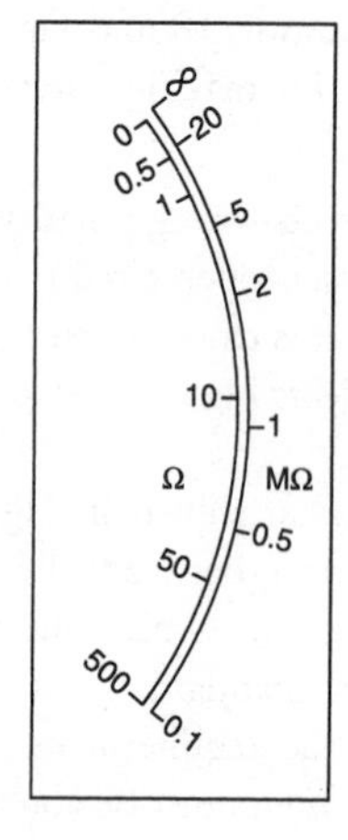

Fig. 6

23. Which one of the following does *not* fall within the detailed inspection of installed electrical equipment?

a no signs of visible damage or defect
b products manufactured to BS5750, Part 1
c satisfactory selection and erection
d the presence of a recognized mark or certificate.

24. What is the minimum distance a 13 A BS1363 socket-outlet can be installed near a shower cubicle or bath?

a 2500 mm
b 2000 mm
c 1750 mm
d 1500 mm.

25. Which one of the following premises only needs periodic inspecting and testing at 10 yearly intervals?

a domestic premises
b church
c leisure complex
d laundrette.

ASSESSMENT 9.3

Answers, Hints and References

1. **b** *See* Ref. 28, no.3, p 41.
2. **b** *See* Ref. 4, section 711.
3. **d** *See* Ref. 18, p49.
4. **c** *See* Ref. 28, no.3, p 99.
5. **a** *See* Ref. 4, regulation 130–10–01.
6. **c** *See* Ref. 18, section 10.
7. **c**.
8. **a**.
9. **d** *See* Regulation 4.
10. **c** The tester's lead can also be put on the lampholder's cpc.
11. **b** A tong tester generally measures current: *See* Ref. 12, p 126.
12. **d** *See* Ref. 4, section 314.
13. **c** *See* Ref. 4, chapter 71.
14. **d** *See* Ref. 28, no.3.
15. **c** *See* Ref. 18, Appendix 9.
16. **b** 8×0.00644.
17. **c** *See* Ref. 4, regulation 412–03–02.
18. **c** $Re = (1 \times 3)/(1 + 3) = 0.75\ \Omega$
19. **b** *See* Ref. 1, p 82, Fig. 4.3.
20. **c** *See* Ref. 1, p 86 and Ref. 4, Table 71A.
21. **d** *See* Ref. 1, p 93.
22. **b** On the continuity resistance range.
23. **b** BS5750 refers to Quality Assurance.
24. **a** *See* Ref. 4, regulation 601–10–03.
25. **a** *See* Ref. 2, p 5, Table 1.1.

10

ASSOCIATED ELECTRONICS TECHNOLOGY

To tackle the assessments in Topic 10 you will need to know:

- special safety precautions associated with electronic maintenance;
- types of electronic circuit diagrams and state their use;
- various types of components used in electronic circuits;
- how to identify common semiconductor devices such as the diode, zener diode, transistor, thyristor, diac, triac, etc.
- how to indicate the values of resistors and capacitors by colour code and letter code;
- how to set up analogue and digital instruments and the cathode ray oscilloscope for testing and measurement;
- how to recognize methods of interconnection such as edge connectors, plugs and sockets, multipin, coaxial, wire wrapping, crimping and insulation displacement;
- how to use hand tools, soldering irons, wire strippers, desoldering tools etc, and also how to carry out connections on circuit boards;
- how to distinguish between alternating and digital waveforms;
- how to describe the action of components in circuits such as resistors, capacitors and listed semiconductor devices including logic gates.

DEFINITIONS

Air-cored inductor – a type of inductor that has no iron magnetic path and, as a consequence, has small inductance. It is used for high frequency work such as in radio tuning circuits.
Amplifier – a device whose output is a magnified function of its input.
Analogue waveform – a waveform that produces a continuous range of values rather than a specific value.
Anode – the positive terminal of a semiconductor device.
Base – the region in a semiconductor transistor situated between emitter and collector.
Binary code – a numerical code consisting of two digits 0 and 1 used for counting in electronic circuits. The digit 0 is low voltage and the digit 1 high voltage.
Capacitor – an electrical component capable of storing charge. In a dc circuit the capacitor blocks current as it becomes fully charged but in an ac circuit current appears to flow continuously through it. NB: An electrolytic capacitor is a polarized capacitor offering high capacitance.
Cathode – the negative terminal of a semiconductor device.
Cathode ray oscilloscope – a high impedance voltmeter designed as a measuring and test instrument for displaying waveforms.
Collector – the region in a semiconductor transistor, normally reverse-biased with respect to the base.
Depletion layer – a region within a semiconductor material where there are no charge carriers.
Desoldering iron – a tool designed to remove solder from a soldered joint, working like a bicycle pump in reverse.
Diac – a triggering semiconductor device with a high turn on current used to extend the range of control in a triac circuit.
Digital waveform – a waveform that shows changes between two definite states that are either on or off.

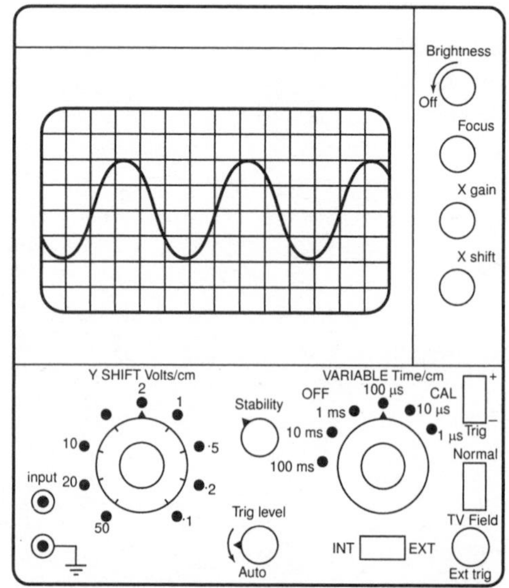

Cathode ray oscilloscope

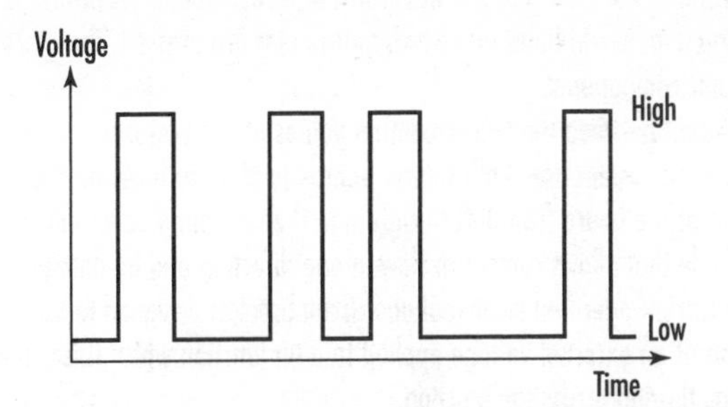

Digital waveform

Diode – a semiconductor pn junction device which can act like a rectifying element when connected to an ac supply. A zener diode is a special diode with a predetermined reverse breakdown voltage and a light emitting diode is one that emits light when current passes between its anode and cathode terminals.

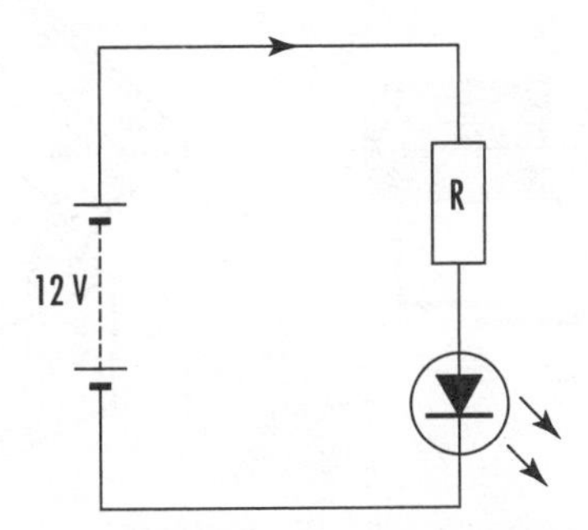

Light emitting diode

Emitter – a region within a semiconductor transistor which is forward-biased with respect to the base.
Forward bias – the application of an external voltage applied to a pn semiconductor junction which reduces the potential barrier in the depletion layer region and allows electrons to flow across the junction.
Frequency – of a waveform, describes the number of repetitive cycles that occur in one second. The duration of one cycle is termed the periodic time.
Fundamental waveform – a sinusoidal waveform having the same frequency as the original waveform.
Harmonic waveform – a sine wave with a frequency that is a multiple of the fundamental and with a smaller amplitude.
Heat sink – a relatively large piece of metal that is placed in contact with a semiconductor device to help dissipate its heat.
Inductor – an electrical component usually in the form of a coil of wire having inductance as its chief property. Ferrite, air-cored and iron-dust inductors are used in high frequency work, e.g. radio tuning, special transformer and computer memories, whereas iron core inductors are used in low frequency work such as fluorescent lamp chokes.
Insulation displacement – a method of terminating flat multicore ribbon cable to various types of cable connector, such as edge connectors and transition connectors etc.
Logic circuit – a circuit that carries out simple logic functions.
Logic gate – a digital device that produces an output of logic 1 or 0 depending upon the combination of inputs.

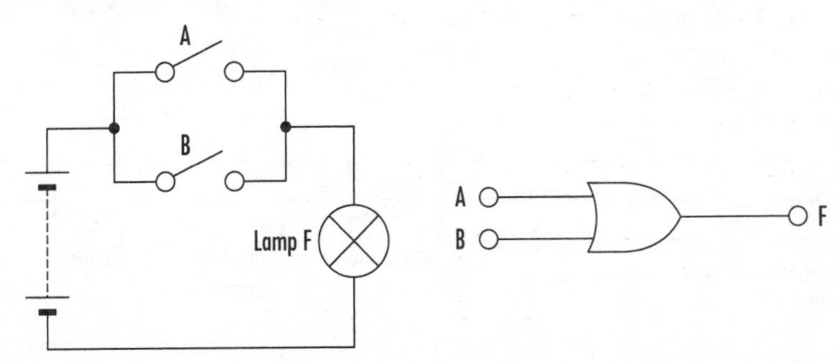

OR gate gate circuit

Mark-to-space ratio – the ratio of the time that the waveform of a rectangular waveform is high to the time that it is low.
Matrix board – a board having a matrix of holes into which matrix pins are pressed. These provide terminal posts for the connection and soldering of electronic components.
Peak-to-peak – the measurement between the two maximum values of a waveform.
Positional reference system – a system used with matrix boards to identify holes on the board. This is achieved by counting along the columns at the top of the board from left to right and then counting down the rows.
Rectifier – a semiconductor diode that allows current to flow in one direction and by doing so can convert ac to dc.
Residual current device – a current-operated earth leakage circuit breaker designed to trip out when it senses an earth fault.
Reverse bias – the application of an external voltage applied to a pn junction which raises the barrier potential in the depletion layer and prevents electrons flowing across the junction.
Stripboard – a board that has rows of copper strip bonded on one side of the board. Connections between components are soldered on the copper strip.

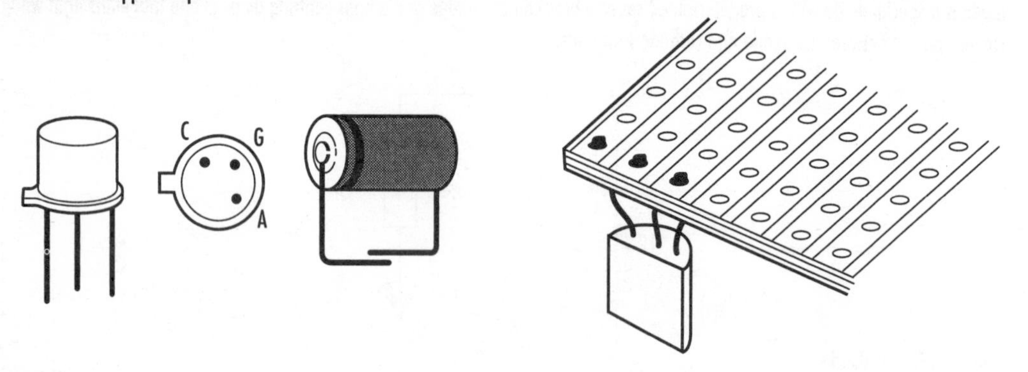

Electronic components **Soldered component on a stripboard**

Square wave – a waveform used in digital circuits in which the mark-to-space ratio is 1. It is used mainly in timing and oscillator circuits.
Thermistor – a semiconductor sensor possessing a high temperature coefficient of resistance and high sensitivity to temperature changes.
Thyristor – a four layer (pnpn) silicon controlled rectifier which has a gate terminal that allows it to be externally pulsed on.
Transformer – an electromagnetic device, mostly used for changing voltage from one level to another (stepping up or stepping down).
Transistor – a semiconductor device that has three terminals and is used for switching and amplification purposes.
Triac – a full-wave semiconductor switching device used in power circuits.

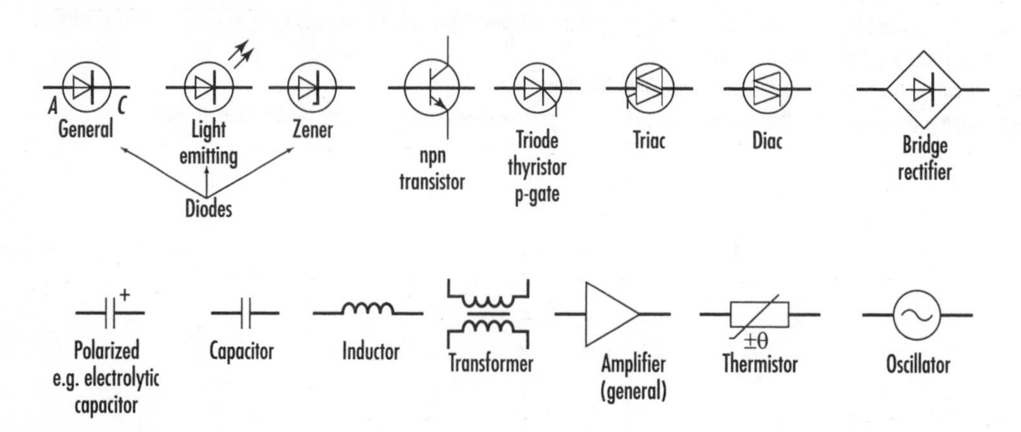

BS3939 circuit symbols

TOPIC 10

Associated Electronics Technology

ASSESSMENTS 10.1 – 10.3

Time allowed: 1 hour

Instructions

* You should have the following:

  Question Paper
  Answer Sheet
  HB pencil
  Metric ruler

* Enter your name and date at the top of the Answer Sheet.
* When you have decided a correct response to a question, on the Answer Sheet, draw a straight line across the appropriate letter using your HB pencil and ruler (see example below).
* If you make a mistake with your answer, change the original line into a cross and then repeat the previous instruction. There is only one answer to each question.
* Do not write on any page of the Question Paper.
* Make sure you read each question carefully and try to answer all the questions in the allotted time.

Example:

a 400 V	**a** 400 V
~~b~~ 315 V	~~b~~ (crossed) 315 V
c 230 V	**c** 230 V
d 110 V	~~d~~ 110 V

ASSESSMENT 10.1

1. A supplementary measure of protection against electric shock when servicing electronic equipment is to use

a a rubber mat over the floor area of working
b a 100 mA tripping residual current device
c a lockable miniature circuit breaker feeding local socket outlet
d only low voltage equipment.

2. The type of electronic component assembly board shown in Fig. 1 is called a

a punch board
b matrix board
c spacer board
d reference board.

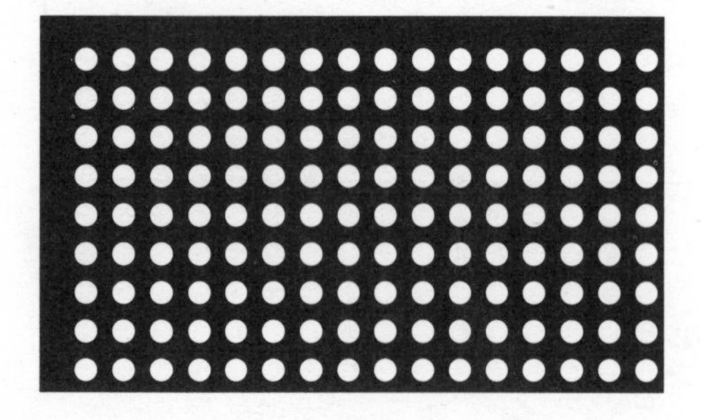

Fig. 1

3. The type of system that is used to identify electronic components on the board in Fig. 1 is called a

a schematic reference system
b soldering layout system
c positional reference system
d grid layout circuit system.

4. What is the name of the BS3939 electronic symbol shown in Fig. 2?

a junction diode
b zener diode
c transistor
d thyristor.

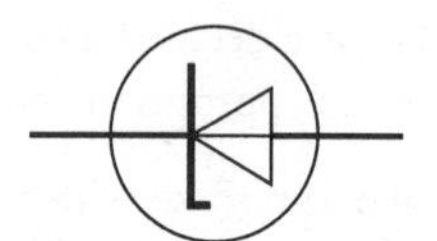

Fig. 2

5. Fig. 3 shows a simple lamp dimmer circuit. The component marked X is called a

a diac
b triac
c thyristor
d transistor.

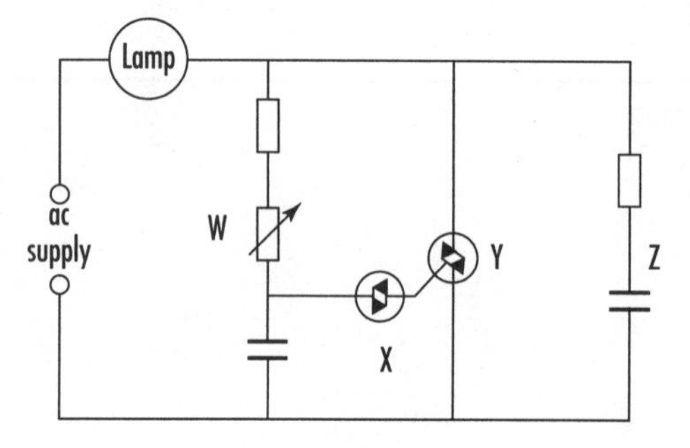

Fig. 3

6. In Fig. 3 the component marked Y is called a

a transistor
b thyristor
c triac
d diac.

7. In Fig. 3 the resistor marked W acts in the circuit like a

a blocking component
b power component
c voltage regulator
d surge arrestor.

8. What is the name given to the electronic components in Fig. 4 making up the seven-segment display?

a low energy diodes
b luminescent diodes
c photoelectric diodes
d light emitting diodes.

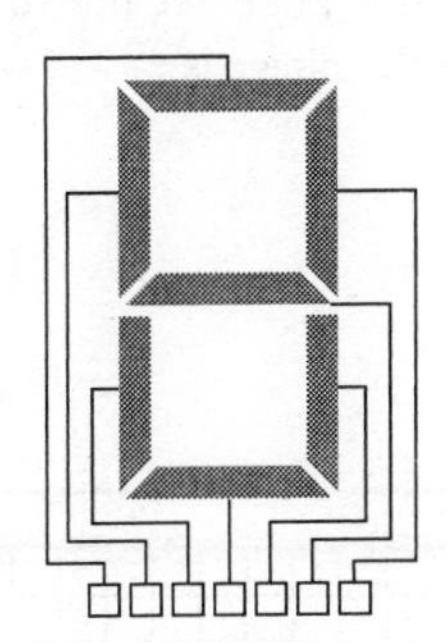

Fig. 4

9. The dot shown on the IC package in Fig. 5 indicates pin number

a 18
b 10
c 9
d 1.

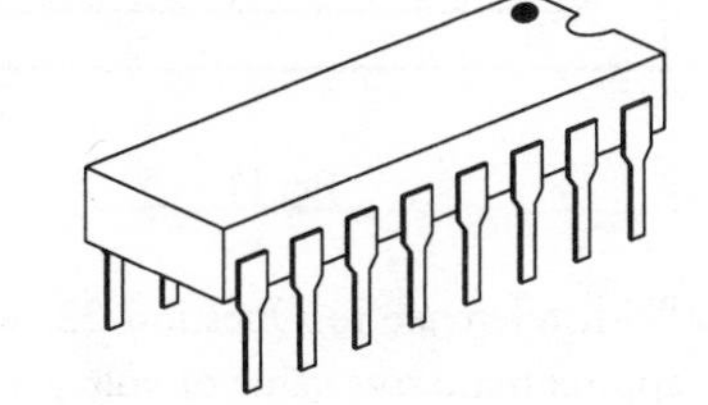

Fig. 5

10. The two rows of pins shown in Fig. 5 are referred to as

a dual-in-line contacts
b 'DIN' contacts
c double-side contacts
d flat pack contacts.

11. Fig. 6 shows the structure of a **pnp transistor**. What is the name given to the terminal marked X?

a receiver
b emitter
c load
d common.

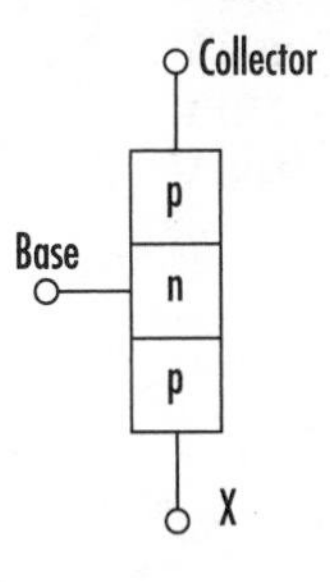

Fig. 6

12. What is the name of the solid state component abbreviated to FET?

a Field Effect Transistor
b Feedback Energy Transmitter
c Frequency Energy Transducer
d Fast Electronic Trigger.

13. What is the name of the device shown in Fig. 7 that is used to control temperature in circuits?

a capacitor
b thermistor
c rheostat
d regulator.

Fig. 7

14. The value of the colour code resistor shown in Fig. 8 is

a 30 kΩ ± 10%
b 27 kΩ ± 5%
c 20 kΩ ± 2%
d 18 kΩ ± 1%.

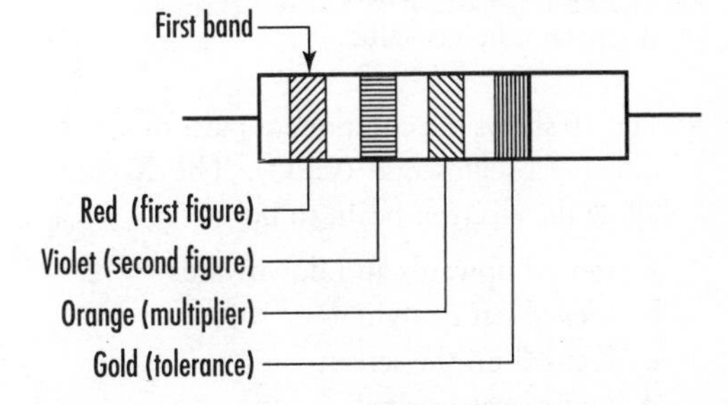

Fig. 8

15. What is the maximum working voltage of the **polyester capacitor** shown in Fig. 9?

a 500 V
b 400 V
c 250 V
d 100 V.

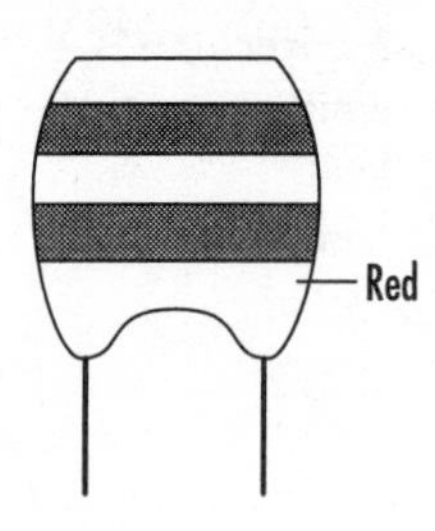

Fig. 9

16. What is the value and tolerance of a resistor identified as R33M?

a 33.0 Ω, ± 05%
b 3.3 Ω, ± 10%
c 0.33 Ω, ± 20%
d 0.30 Ω, ± 30%.

17. What is the value of a capacitor marked 4p5?

a 4.5×10^{-12} F
b 4.5×10^{-9} F
c 4.5×10^{-6} F
d 4.5×10^{-3} F.

18. When a **harmonic waveform** mixes with the **fundamental waveform** it creates a

a sinusoidal waveform
b complex waveform
c basic waveform
d cyclic waveform.

19. In a sinusoidal waveform, 0.707 times the maximum value is the

a root mean square value
b average value
c peak-to-peak value
d instantaneous value.

20. Fig. 10 shows the component parts of a cathode ray oscilloscope (CRO). The X-plates allow the electron beam to be

a moved upwards and downwards
b moved left or right
c focused on the screen
d made large or small.

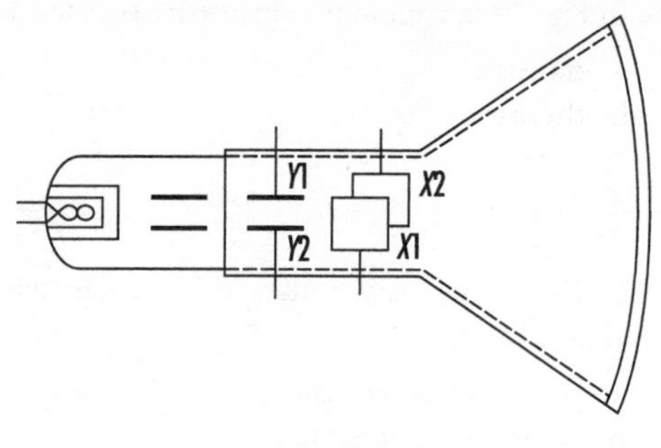

Fig. 10

21. The CRO in Fig. 10 can be used to measure all the following, *except*

a voltage
b amplitude
c frequency
d impedance.

22. Fig. 11 shows a trace on a CRO that has its Y-amp control knob set on 10 V/cm. The **peak-to-peak** voltage is

a 480 V
b 240 V
c 80 V
d 40 V.

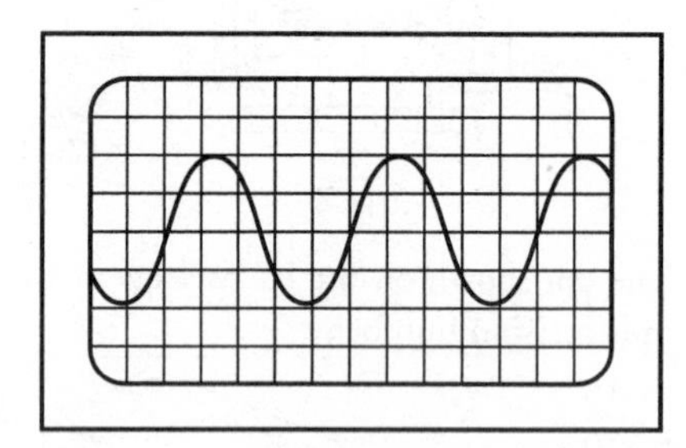

Fig. 11

23. With reference to Question 22, what is the approximate rms value of voltage?

a 240 V
b 40 V
c 28.28 V
d 14.14 V

24. In Fig. 12, any of the positions marked with a dot are called

- **a** reference values
- **b** instantaneous values
- **c** maximum or peak values
- **d** root mean square values.

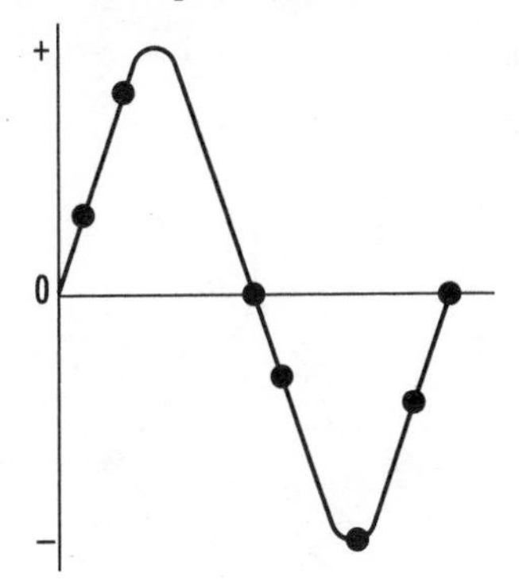

Fig. 12

25. Which one of the following methods of finding a resistor's value considers its tolerance?

- **a** ohmmeter
- **b** multimeter
- **c** ammeter and voltmeter
- **d** colour code.

26. When testing a **diode** with a multimeter on its resistance range, the diode's forward direction should read

- **a** infinity
- **b** between 1 MΩ and 10 MΩ
- **c** between 100 Ω and 1 MΩ
- **d** a low resistance.

27. The crocodile clip shown in Fig. 13 is used as a

- **a** test prod
- **b** temporary fix
- **c** component support
- **d** heat shunt.

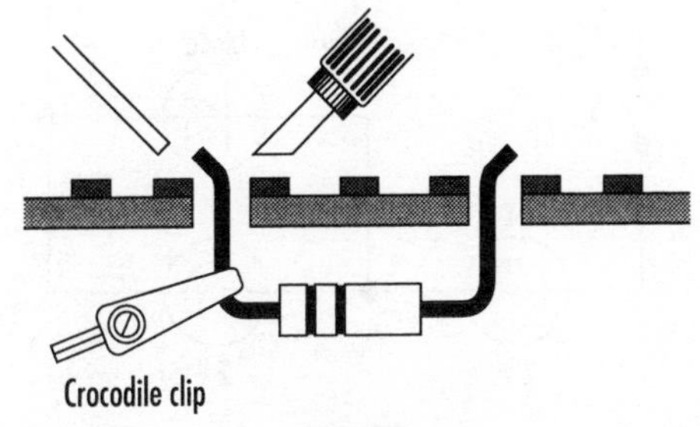

Fig. 13

28. All of the following are important factors to be considered when fault finding in electronic wiring circuits, *except*

- **a** checking supply availability and correct voltage
- **b** checking input and output voltages of individual components
- **c** checking the frequency at the main supply position
- **d** checking to see if soldered joints and connections are sound.

29. The solder used in electronic work consists of

- **a** 60% tin and 40% copper
- **b** 40% tin and 60% lead
- **c** 60% lead and 40% steel
- **d** 40% lead and 60% tin.

30. Multicore solder contains flux which is used to

- **a** remove the oxide film from the surfaces of the metal joint
- **b** increase the conductivity of the metal surfaces being joined
- **c** increase the temperature of the metal surfaces being joined
- **d** melt surplus insulation left close to the surfaces being joined.

31. The item marked X on the soldering iron in Fig. 14 is called a

- **a** shaft
- **b** flute
- **c** bit
- **d** tip.

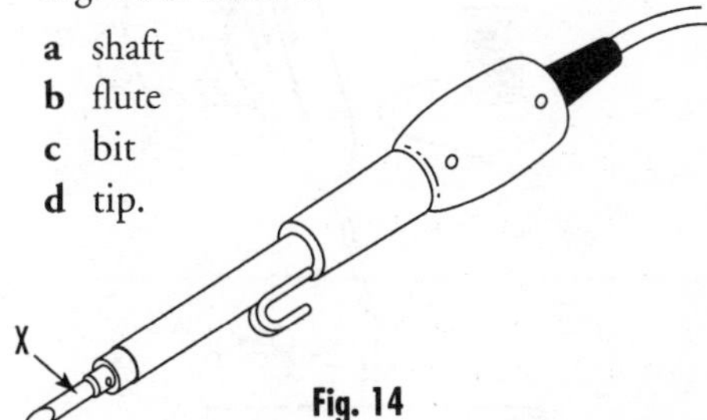

Fig. 14

32. If a soldering iron's temperature is too low or the joint to be soldered is dirty, it might result in a

- **a** porous joint
- **b** soft joint
- **c** wet joint
- **d** dry joint.

33. Fig. 15 shows the method of making a soldered joint to a

a U terminal
b tag terminal
c circuit terminal
d flat-pack terminal.

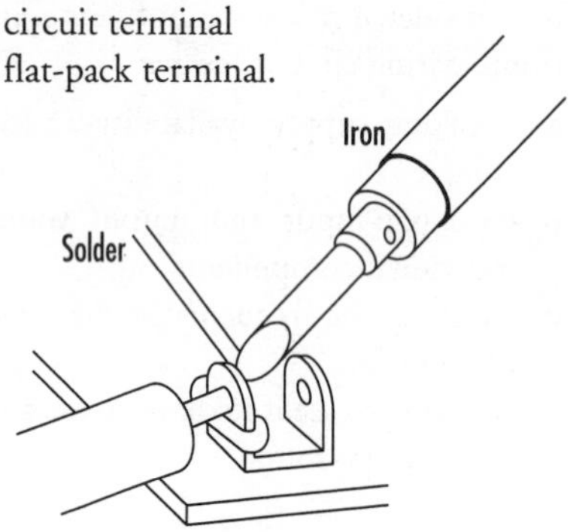

Fig. 15

34. With reference to Fig. 15, when the soldering process is nearing completion, you should

a remove the iron before the solder
b remove the solder when it smokes
c remove the solder before the iron
d remove the iron and solder together.

35. The portable tool used in Fig. 16 is called a

a wire wrapping tool
b soldering iron gun
c solderless tool
d pin terminal tool.

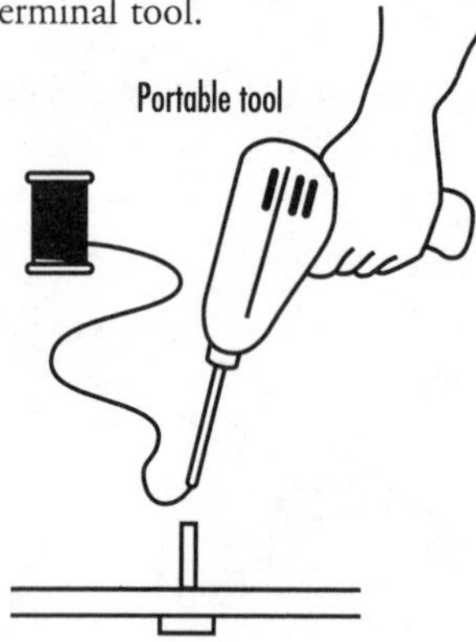

Fig. 16

36. Which one of the following is the correct tool to use to remove a faulty component soldered on a circuit board?

a Stanley knife
b junior hacksaw
c long-nosed pliers
d desoldering iron.

37. Which letter in Fig. 17 identifies the positive dc load connection?

a W
b X
c Y
d Z.

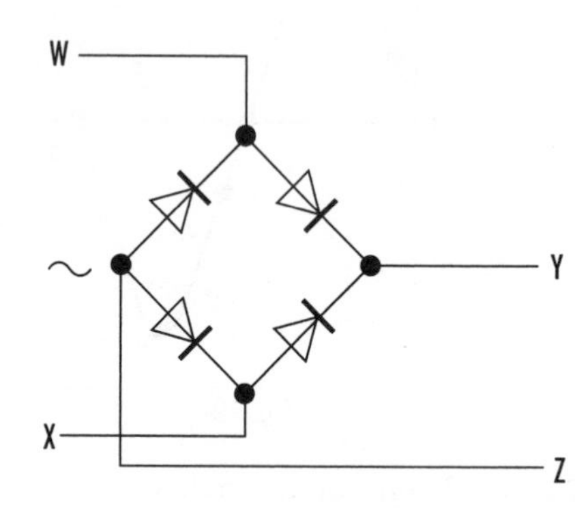

Fig. 17

38. The technique called **insulation displacement** is a method of terminating

a flat T-ribbon cables
b circular PVC sheathed cables
c jack connectors
d DIN-style audio connectors.

39. In a pn junction diode, an increased **depletion layer** is the result of

a forward bias
b reverse bias
c acceptor ions
d charge carriers.

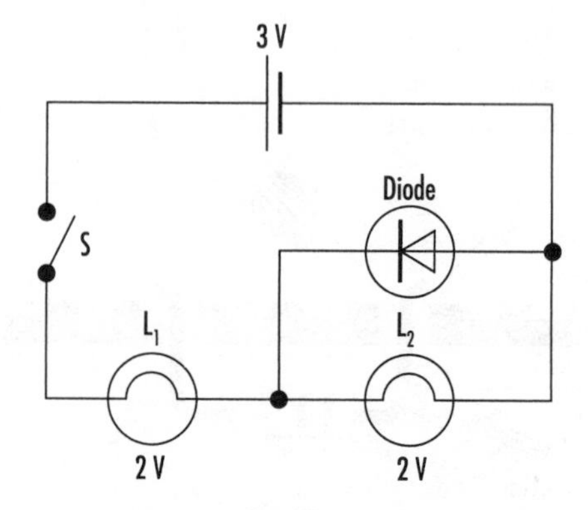

Fig. 18

40. In Fig. 18, when the switch marked S is closed, what is the condition of each lamp?

a both lamps will be bright
b L_2 will be bright and L_1 dim
c L_1 will be bright and L_2 dim
d both lamps will be dim.

41. What is the term that expresses time taken for a sine wave to make one complete oscillation?

a period
b frequency
c wavelength
d attenuation.

42. All of the following are designed to operate using an **analogue waveform,** *except* a

a radio
b television
c telephone
d calculator.

43. The type of wave shown in Fig. 19 is called a

a block waveform
b digital waveform
c peak waveform
d analogue waveform.

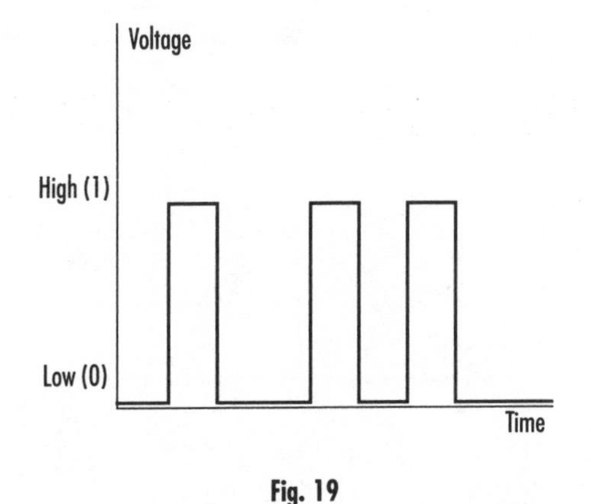

Fig. 19

44. The **mark-to-space ratio** of the rectangular wave shown in Fig. 20 is

a 4
b 3
c 2
d 1.

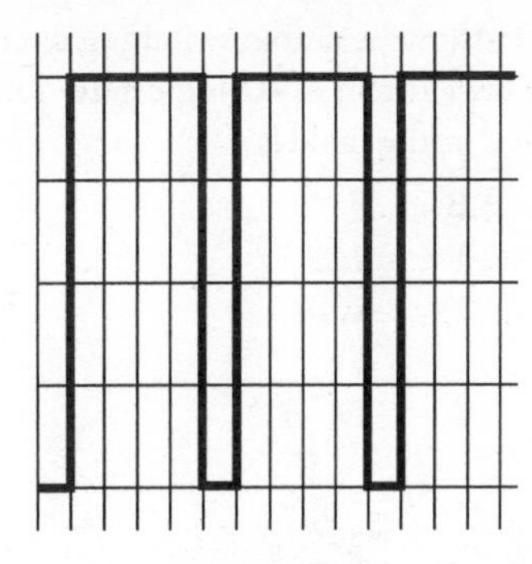

Fig. 20

45. To make a **thyristor** operate, a pulse is needed at its

a gate terminal
b base terminal
c emitter terminal
d collector terminal.

46. The semiconductor device shown in Fig. 21 emits

a light
b heat
c magnetism
d sound.

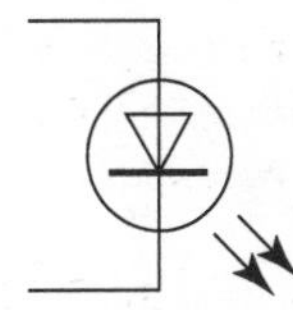

Fig. 21

47. The circuit arrangement shown in Fig. 22. is designed to give

a voltage transformation
b half-wave rectification
c full-wave rectification
d low voltage regulation.

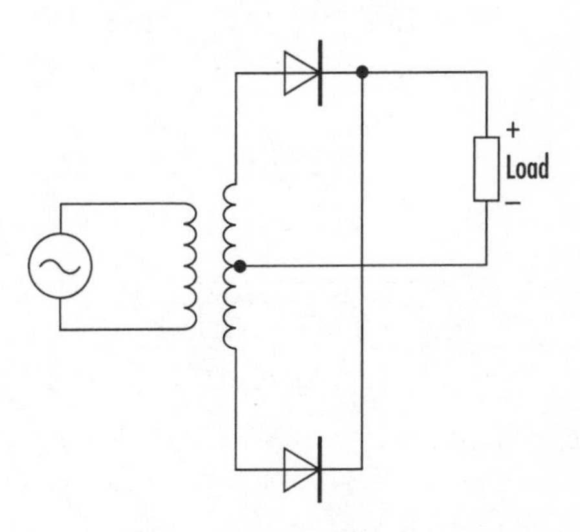

Fig. 22

48. Fig. 23 shows a simple switching circuit and truth table for an **AND logic gate**. The missing line in the table is

	A	B	F
a	1	1	0
b	0	1	1
c	1	1	1
d	1	0	1.

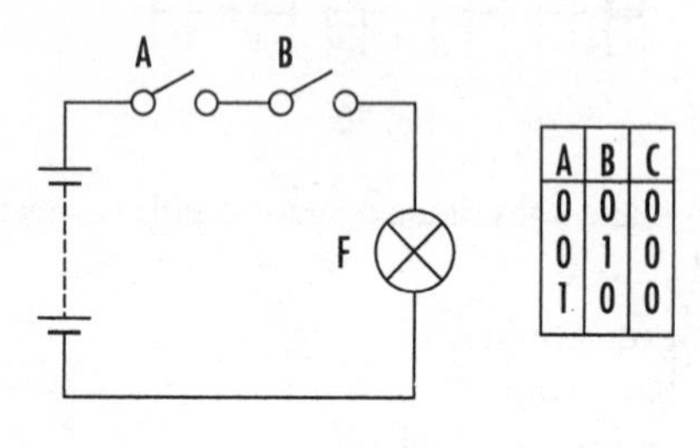

A	B	C
0	0	0
0	1	0
1	0	0

Fig. 23

49. In a simple **OR logic gate** the operating function can be simulated by a lamp and dc source connected to

a one two-way switch
b two one-way switches in series
c two one-way switches in parallel
d one intermediate switch.

50. All of the following are types of logic gates, *except*

a NOT
b NOR
c NAND
d NORT.

ASSESSMENT 10.1

Answers, Hints and References

1. **a** Alternatively 30 mA red protection.
2. **b** *See* Ref. 27, p 24, Fig. 2.15.
3. **c** *See* Ref. 27, p 25, Fig. 2.18.
4. **b** *See* Ref. 6, pp 61–62.
5. **a** *See* Ref. 6, p 65, Fig. 4.25.
6. **c** *See* page references given in answers to questions 5 and 7.
7. **c** *See* Ref. 27, p 105, Fig. 7.23.
8. **d** *See* Ref. 27, pp 84–85.
9. **d** *See* Ref. 27, p 89, Fig. 6.20.
10. **a** *See* Ref. 27, p 89.
11. **b** *See* Ref. 27, p 87, Fig. 6.16.
12. **a** *See* Ref. 27, p 86.
13. **b** *See* Ref. 27, pp 85–86.
14. **b** *See* Ref. 6, p 4, Fig. 1.1.
15. **c** *See* Ref. 27, p 7.
16. **c** *See* Ref. 7, p 14.
17. **a** *See* Ref. 6, p 3, Table 3.
18. **b** *See* Ref. 6, p 68, Fig. 4.32.
19. **a** *See* Ref. 7, p 13.
20. **b** *See* Ref. 3, p 109, Fig. 6.3.
21. **d**.
22. **d** *See* Ref. 27, pp 72–73.
23. **d** $20 \times 0.707 = 14.14$ V.
24. **b** *See* Ref. 6, p 11.
25. **d**.
26. **d**.
27. **d** *See* Ref. 27, p 23, Fig. 2.12.
28. **c** Frequency is checked on equipment.
29. **d**.
30. **a** *See* Ref. 27, p 20.
31. **c** *See* Ref. 27, p 21, Fig. 2.6.
32. **d** *See* Ref. 27, p 21.
33. **b** *See* Ref. 27, p 22, Fig. 2.8.
34. **c** *See* Ref. 27, p 21.
35. **a** *See* Ref. 27, p 26.
36. **d** *See* Ref. 27, pp 22–23.
37. **c** *See* Ref. 27, p 96, Fig. 7.6.
38. **a** *See* Ref. 27, p 28, Fig. 2.23.
39. **b** *See* Ref. 6, pp 60–61.
40. **d** The diode blocks current flow.
41. **a** $T = 1/f$. If the supply frequency is 50 Hz, each complete oscillation takes 1/50 second (0.02 s).
42. **d** *See* Ref. 27, p 69.
43. **b** *See* Ref. 6, p 66, Fig. 4.28.
44. **a** *See* Ref. 6, p 68.
45. **a** *See* Ref. 27, p 102, Fig. 7.17.
46. **a** *See* Ref. 27, p 84, Fig. 6.9.
47. **c** *See* Ref. 7, p 99, Fig. 5.6.
48. **c** *See* Ref. 27, p 116, Fig. 8.2.
49. **c** *See* Ref. 27, p 116, Fig. 8.3.
50. **d**.

ASSESSMENT 10.2

1. A protective measure used to supplement safety when servicing electronic equipment is to use a residual current device designed with a tripping sensitivity of

a 300 mA
b 100 mA
c 50 mA
d 30 mA.

2. Before checking any faulty electronic equipment connected to the supply, it is important to

a isolate the equipment and make visual tests
b see if the equipment has its own built–in fuse
c make a second supply test on the equipment
d identify the equipment's supply voltage.

3. If the lead of a soldering iron becomes damaged while in use, you should

a keep the damaged part away from exposed metalwork
b immediately assess the damage and repair it
c immediately remove the soldering iron from service
d continue to use the soldering iron with caution.

4. Which one of the following types of diagram shows the physical interconnections of electronic components?

a schematic wiring diagram
b single line block diagram
c internal circuit diagram
d positional reference diagram.

5. Continually heating and applying extra solder, to improve a soldered joint in a cable, will result in its

a conductivity becoming too high
b appearance becoming dull and spiky
c resistance becoming too high
d insulation becoming damaged.

6. The main hazard encountered when using a soldering iron in a **confined space** is

a flying fragments of solder
b burns from the soldering iron
c fumes from the soldering flux
d insufficient access to solder.

7. What is the name of the type of tool shown in Fig. 1, used for removing solder?

a decompression iron
b desoldering iron
c solder suction iron
d solderless iron.

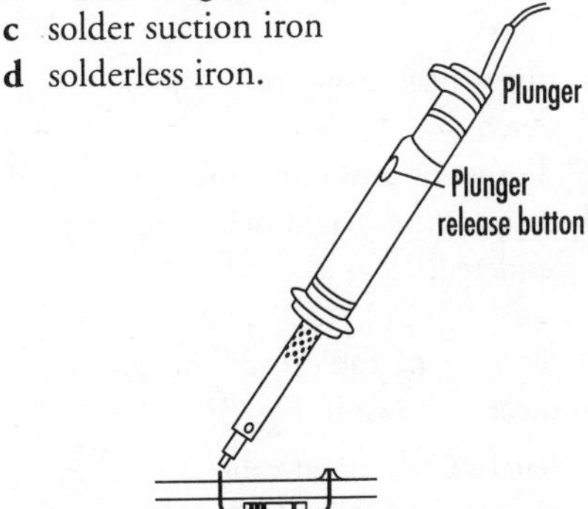

Fig. 1

8. The type of electronic component assembly board shown in Fig. 2 is called a

a stripboard
b breadboard
c circuit board
d reference board.

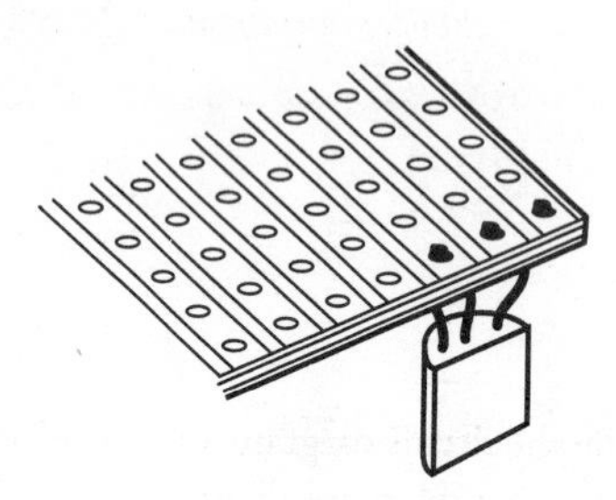

Fig. 2

9. What is the name given to the grid system in Fig. 2 that is used to wire electronic components?

a schematic reference system
b soldering layout system
c positional reference system
d grid layout circuit system.

10. The phase conductor in a three-core flexible cord is identified by the colour

a brown
b cream
c blue
d red.

11. A heat shunt is used in electronic assembly work to

a quickly cool a soldered joint or a component
b divert heat away from a joint being soldered
c dispose of unwanted amounts of solder
d support the weight of a component being soldered.

12. What type of cable uses the indirect edge connector shown in Fig. 3?

a flat PVC sheathed cable
b circular PVC sheathed cable
c PVC sheathed screen cable
d flat T-ribbon PVC cable.

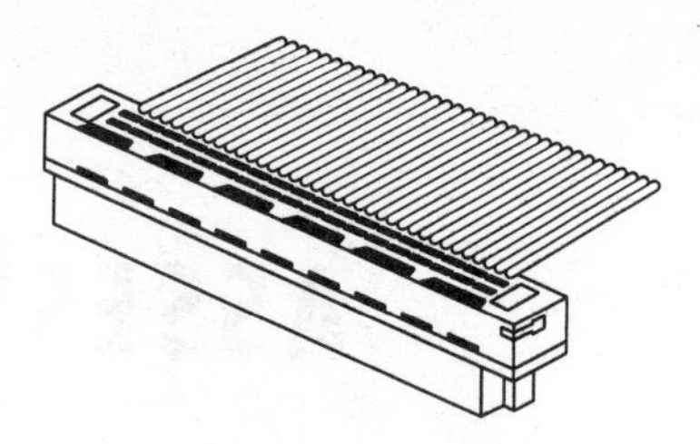

Fig. 3

13. All of the following methods can be used to determine the ohmic value of a resistor, *except*

a a low-high reading ohmmeter
b a proprietary multimeter
c an ammeter and voltmeter
d a continuous ringing bell.

14. Fig. 4 shows a multimeter with its red terminal connected to the band end of a **pn junction diode**. The band denotes the diode's

a anode connection
b cathode connection
c neutral connection
d positive connection.

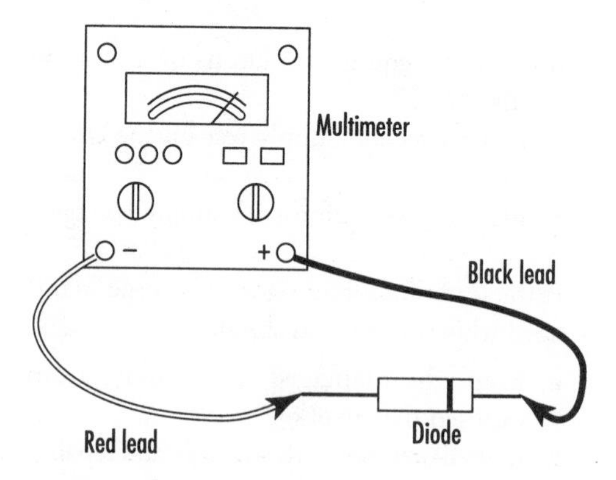

Fig. 4

15. In a diode, the term **reverse bias** means that its current flow is

a blocked
b possible
c bidirectional
d zero.

16. In Fig. 5, which symbol represents a **polarized electrolytic capacitor**?

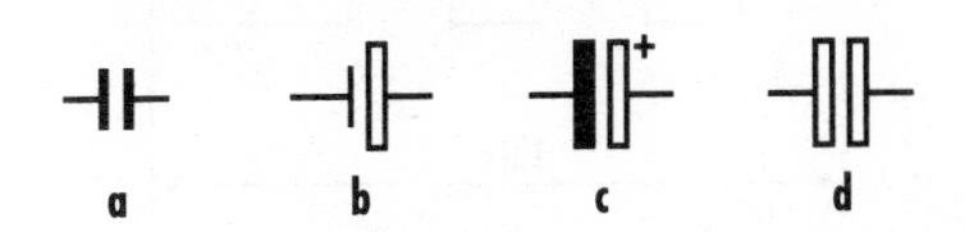

Fig. 5

17. The three connections on a **semiconductor thyristor** are called

a anode, cathode and gate
b base, emitter and collector
c input, output and common
d source, drain and gate.

18. In Fig. 6, what is the name of the component marked X?

a triac
b diac
c transistor
d zener diode.

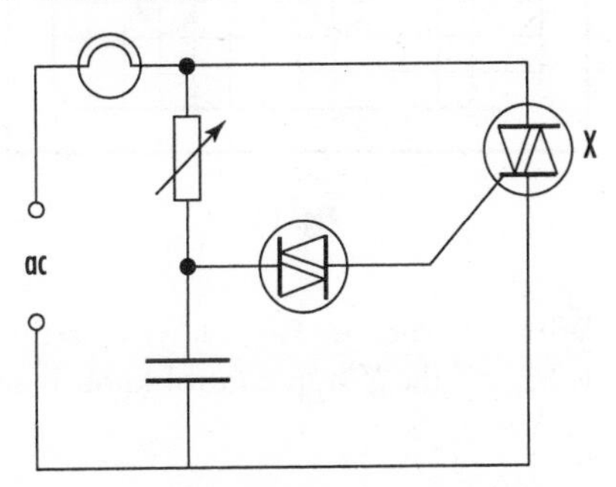

Fig. 6

19. When a resistor is identified with an alphanumeric code the letter R is used as the decimal point or the

a term resistance
b base value in ohms
c tolerance value
d zero.

20. Using the resistor colour code, the colours yellow, blue, brown, gold are identified by bands 1, 2, 3 and 4 respectively. The value of the resistor would be:

a 4.6 MΩ + 10%
b 4600 Ω + 20%
c 460 Ω + 5%
d 4.6 Ω + 1%.

21. What is the value of a code resistor marked 4M7?

a 470 MΩ
b 47 MΩ
c 4.70 MΩ
d 0.47 MΩ.

22. A **triac** is used in a lamp dimming circuit to provide

a amplification
b power control
c rectification
d smoothing.

23. What is the component positional reference of 'TR1' in Fig. 7?

a F10
b E11
c F11
d G11.

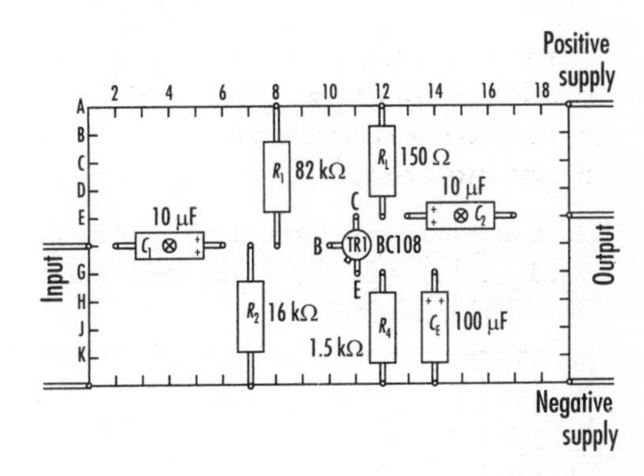

Fig. 7

24. Which one of the following describes the **analogue waveform** of the ac supply mains?

a sinusoidal
b sawtooth
c rectangular
d complex.

25. In a **pn junction** diode, p-type material forms the diode's

a cathode polarity
b anode polarity
c neutral polarity
d negative polarity.

26. The complex waveshape shown in Fig. 8 is created by the **fundamental wave** and

a second harmonic
b third harmonic
c fifth harmonic
d seventh harmonic.

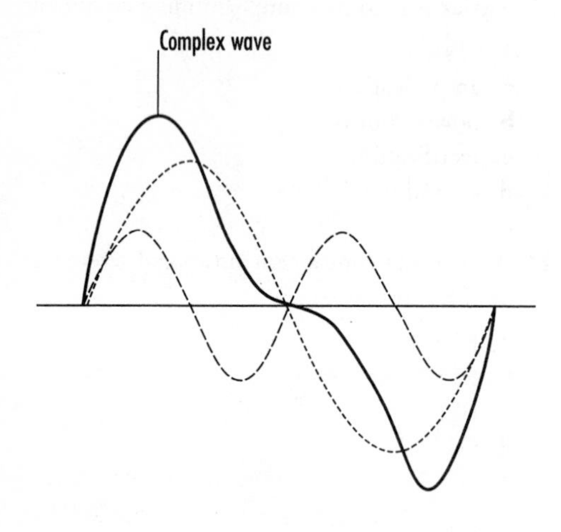

Fig. 8

27. In a sinusoidal waveform, if the root mean square value was 240 V, the peak value would be approximately

a 377 V
b 339 V
c 250 V
d 120 V.

28. In Fig. 9 which one of the waveshapes is a digital signal?

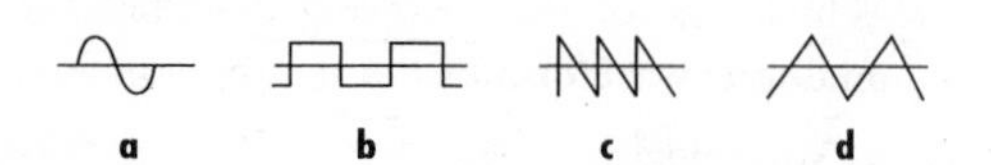

Fig. 9

29. The frequency of the signal shown in Fig. 10 is

a 50.0 Hz
b 8.0 Hz
c 3.5 Hz
d 1.0 Hz.

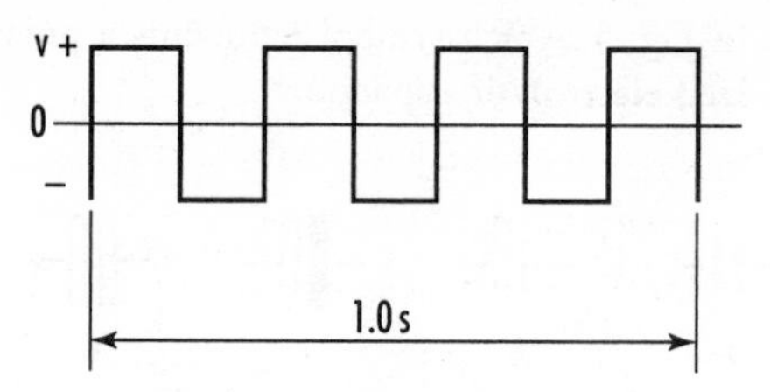

Fig. 10

30. Fig. 11 shows a trace on a CRO that has its time-base control knob set at 2 ms/cm. The frequency is calculated to be

a 100 Hz
b 50 Hz
c 20 Hz
d 10 Hz.

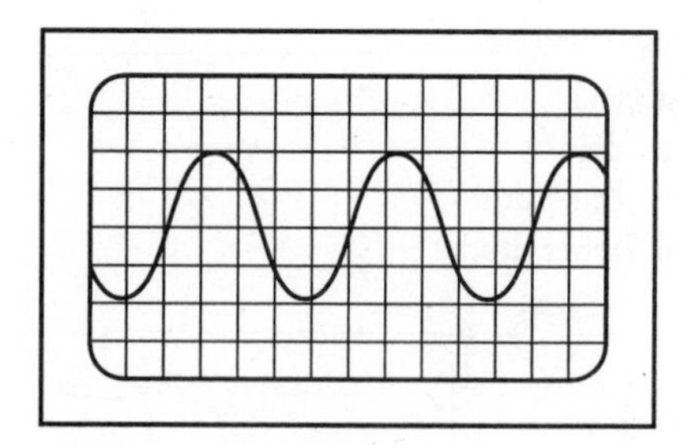

Fig. 11

31. With reference to Fig. 11 what is the rms voltage if the Y-amp control knob is set on 10 V/cm?

a 240 V
b 40 V
c 20 V
d 14.14 V.

32. With reference to Fig. 11, any measured value on the waveform is called

a a reference value
b an instantaneous value
c a peak value
d an actual value.

33. All the following methods can be used to calculate the ohmic value of a resistor, *except*

a a wattmeter and ammeter
b an ammeter and frequency meter
c an ammeter and voltmeter
d a wattmeter and voltmeter.

34. When testing a **diode** with a multimeter on its resistance range, one direction showed that it had a low ohmic reading. This would be the connection for

a blocking ac
b rectifying dc
c reverse bias
d forward bias.

35. Which one of the following is used as a semiconductor material?

a germanium
b copper
c gold
d brass.

36. A diode may be used as a rectifying component because it

a offers electric shock risk protection
b allows low level current to flow
c allows current flow in one direction
d can be used on ac and dc supplies.

37. A **thyristor** is a semiconductor device used for

a sensing high temperatures
b smoothing circuits
c controlled rectification
d voltage amplification.

38. The BS3939 symbol shown in Fig. 12 represents

a a ferrite inductor
b an air-cored inductor
c an iron-cored inductor
d a non-resistive inductor.

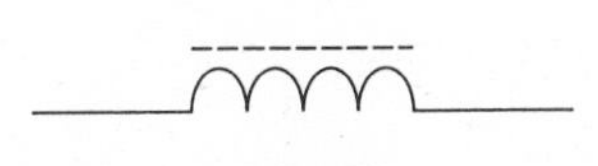

Fig. 12

39. Fig. 13 shows four pn diodes of a bridge-connected rectifier. To which two terminals would you connect the ac supply?

a 1 and 3
b 2 and 4
c 3 and 5
d 4 and 1.

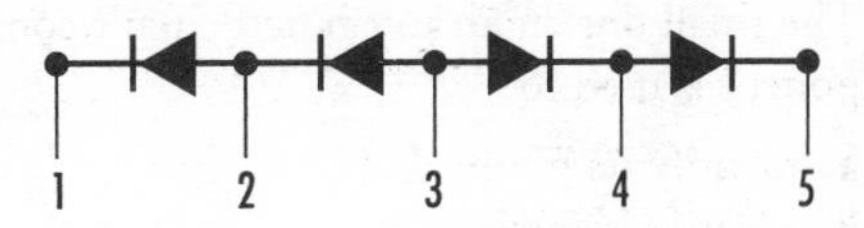

Fig. 13

40. Which two terminals in Fig. 13 need joining together to provide the positive dc output?

a 1 and 2
b 1 and 3
c 1 and 4
d 1 and 5.

41. On which terminal in Fig. 13 would you connect the dc negative output?

a 1
b 2
c 3
d 4.

42. Which one of the following BS3939 symbols in Fig. 14 is a light-sensitive diode?

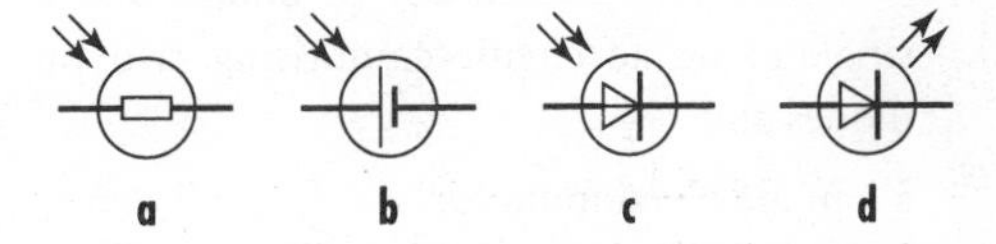

Fig. 14

43. Fig. 15 shows the method of making a soldered joint to a

a U terminal
b tag terminal
c circuit terminal
d pin terminal.

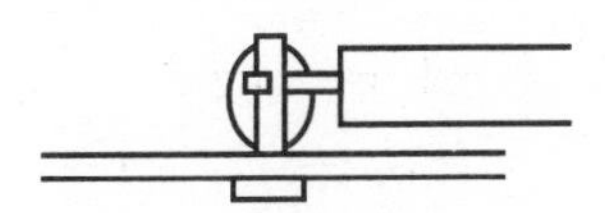

Fig. 15

44. When a soldering iron is being used with an iron-plated bit, before being tinned the bit should be cleaned with a

a damp sponge
b damp woollen cloth
c smooth file
d turps substitute.

45. The small dot on an integrated circuit component is used to

a identify its pin no.1
b identify its type
c indicate its removal
d indicate its position.

46. What is the name of the component shown in Fig. 16?

a tubular inductor
b electrolytic capacitor
c ceramic tube resistor
d coaxial termination.

Fig. 16

47. Full-wave rectification can be obtained in a single-phase ac circuit comprising two pn diodes and

a an air-cored inductor
b a centre-tapped transformer
c an electrolytic capacitor
d a variable resistor.

48. Fig. 17 shows an equivalent switching circuit for a simple

a NOT gate
b OR gate
c NOR gate
d NAND gate.

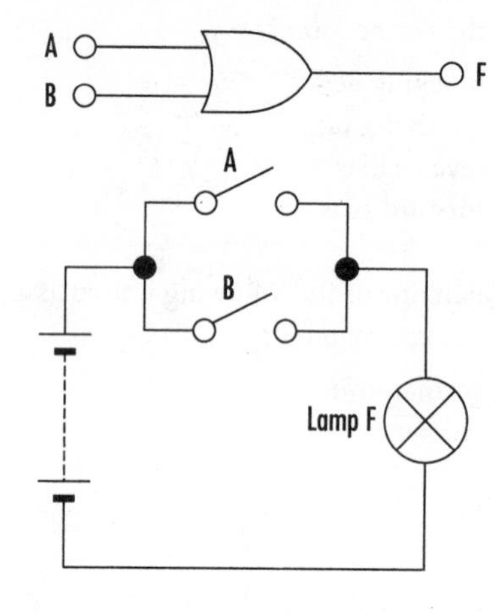

Fig. 17

49. In logic networks, the table that provides a list of all possible binary input signals is called a

a digital table
b logic table
c gate table
d truth table.

50. The type of signal that responds to a continuous range of values rather than a specific value is called

a an analogue signal
b a digital signal
c a pulse signal
d a trigger signal.

ASSESSMENT 10.2

Answers, Hints and References

1. **d** *See* Ref. 27, p 16.
2. **a** *See* Ref. 27, p 29.
3. **b**.
4. **d** *See* Ref. 27, pp 24–25.
5. **d** *See* Ref. 27, pp 20–21.
6. **c**.
7. **b** *See* Ref. 19, p 26, Fig. 3.7.
8. **a** *See* Ref. 27, p 25, Fig. 2.18.
9. **c** *See* Ref. 27, p 24, Fig. 2.17.
10. **a** *See* Ref. 4, Table 51B.
11. **b** *See* Ref. 27, p 23, Fig. 2.12.
12. **d** *See* Ref. 19, p 57, Fig. 3.3.
13. **d**.
14. **b** *See* Ref. 6, p 61, Fig. 4.14.
15. **a** *See* Ref. 7, p 96.
16. **c** *See* Ref. 27, p 6, Fig. 1.6.
17. **a** *See* Ref. 7, p 101, Fig. 5.24.
18. **a** *See* Ref. 7, p 103, Fig. 5.28.
19. **b** *See* Ref. 6, p 14.
20. **c** *See* Ref. 27, p 4, Table 1.1.
21. **c** *See* Ref. 6, p 5.
22. **b** *See* Ref. 27, p 91.
23. **c** *See* Ref. 27, p 25.
24. **a** *See* Ref. 6, p 68.
25. **b** *See* Ref. 6, p 59.
26. **a** *See* Ref. 6, p 68.
27. **b** *See* Ref. 6, p 67.
28. **b** *See* Ref. 6, p 68.
29. **c** *See* Ref. 7, p 99, Fig. 5.6.
30. **a** *See* Ref. 27, pp 73–74.
31. **d** *See* Ref. 27, pp 73–74.
32. **b** *See* Ref. 6, p 11.
33. **b**.
34. **d** *See* Ref. 7, p 96.
35. **a** *See* Ref. 7, p 94.
36. **c** *See* Ref. 7, p 98.
37. **c** *See* Ref. 7, p 100.
38. **a**.
39. **b** *See* Ref. 7, p 99, Fig. 5.18.
40. **d** *See* Ref. 7, p 99, Fig. 5.18.
41. **c** *See* Ref. 7, p 99, Fig. 5.18.
42. **c** *See* Ref. 6, p 13.
43. **d** *See* Ref. 19, Fig. 3.1.
44. **a** *See* Ref. 27, p 20.
45. **a** *See* Ref. 6, p 66, Fig. 4.27.
46. **b** *See* Ref. 27, p 6, Fig. 1.6.
47. **b** *See* Ref. 7, p 99, Fig. 5.16.
48. **b** *See* Ref. 27, p 116, Fig. 8.3.
49. **d** *See* Ref. 27, p 190.
50. **a** *See* Ref. 27, p 184.

ASSESSMENT 10.3

1. When servicing electronic equipment, all of the following protective measures can be used *except*

a a 30 mA residual current device
b an isolating transformer
c placing out of reach
d an insulated bench surface.

2. The purpose of using a **positional reference system** in electronic wiring is to

a inspect component soldering work
b identify holes on a matrix board
c inspect and test connections
d identify component values.

3. Which one of the following electronic components is likely to be connected with a ribbon cable connector?

a stereo music system
b signal generator
c laser-jet printer
d cathode ray oscilloscope.

4. The purpose of using a copper desoldering braid is to

a remove heat from a circuit board component
b remove excess solder from a connection
c reduce the surface temperature of a circuit board
d wipe clean a soldering iron bit.

5. In electronic wiring, the difference between a plain matrix board and a stripboard is that the latter

a is made with copper tracks
b requires wire interconnectors
c cannot be re-soldered
d requires a circuit diagram.

6. What is the name of the silicon controlled rectifier shown in Fig. 1?

a triac
b diac
c thyristor
d transistor.

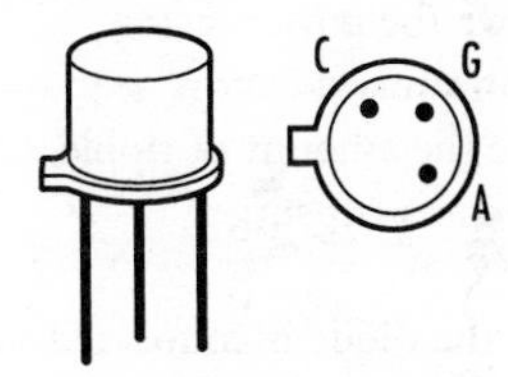

Fig. 1

7. In Fig. 1, the terminal marked C is called the

a common connection
b cathode connection
c chassis connection
d closed connection.

8. A resistor marked 2K9 would have a resistance value of

a 2900 Ω
b 290 Ω
c 29 Ω
d 2.9 Ω.

9. In Question 8 above, if the resistor had a tolerance of 5%, its highest working value would be

a 3045 Ω
b 304.5 Ω
c 30.45 Ω
d 3.045 Ω.

10. Fig. 2 shows the symbol for an npn transistor. What is the unidentified connection lead indicated by the arrow?

a gain
b frame
c gate
d emitter.

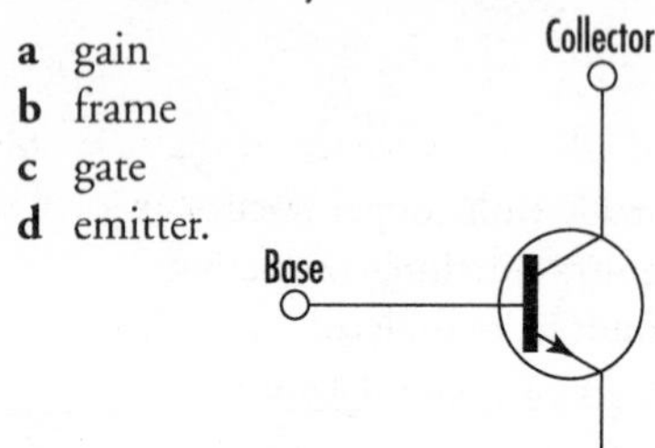

Fig. 2

11. A smoothing capacitor is inserted into the output of a full-wave rectifier in order to

a improve the power factor
b contain unused energy
c reduce the amount of ripple
d avoid transient surges.

12. In Fig. 3 the diode is connected so that

a both lamps will be dim
b L_1 will be bright and L_2 dim
c L_1 will be dim and L_2 bright
d both lamps will not light.

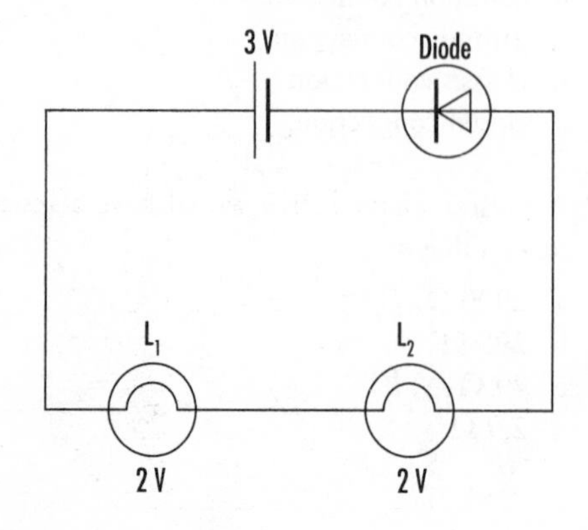

Fig. 3

13. With reference to Fig. 3, the diode's behaviour, in relation to the supply source, is

a reverse biased
b forward biased
c semiconducting
d discharging.

14. The purpose of the zener diode in Fig. 4. is to provide

a stabilized output voltage
b negative feedback
c forward bias
d circuit protection.

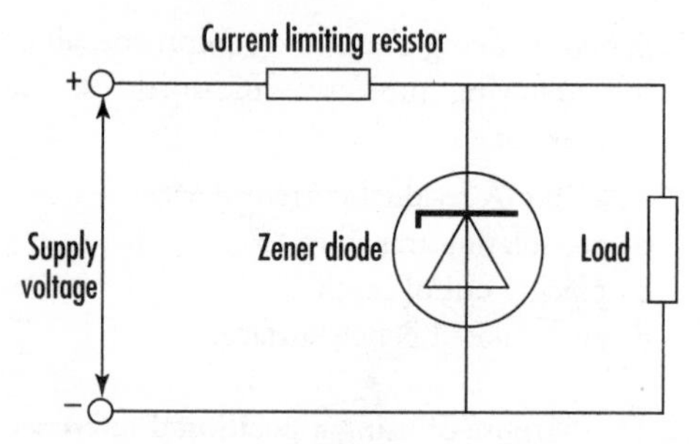

Fig. 4

15. Which one of the following electronic components develops a relatively large internal **depletion layer** when connected to oppose the flow of current?

a air-core inductor
b polyester capacitor
c pn diode
d carbon resistor.

16. Fig. 5 shows the front view of a simple cathode ray oscilloscope. When this is connected in a circuit the trace on the screen is created by small particles called

a ions
b neutrons
c protons
d electrons.

17. With reference to Fig. 5, vertical and horizontal movement of the trace is achieved by means of an internal

a gun
b accelerating anode
c deflection system
d control grid.

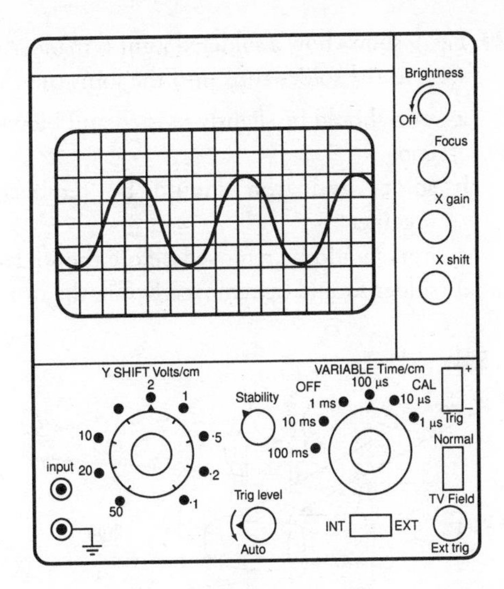

Fig. 5

18. With reference to Fig. 5, to measure the voltage of the trace on the screen, the user needs to operate the control knob marked

a time/cm
b volts/cm
c trig level
d stability.

19. Which one of the following output signals in Fig. 6 would represent the waveshape for full-wave rectification?

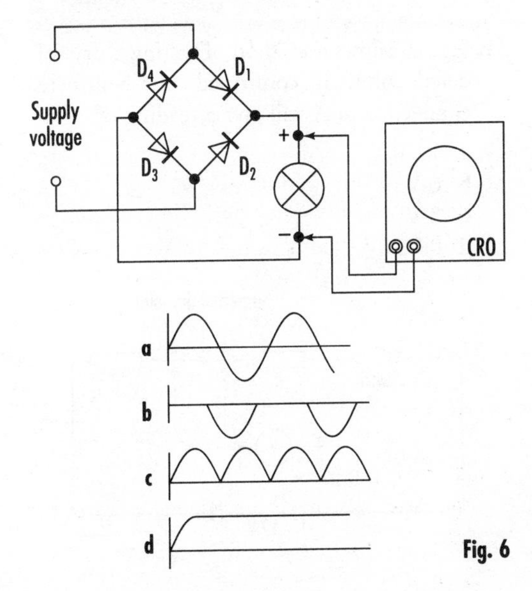

Fig. 6

20. Which one of the following waveforms in Fig. 7 is a digital signal?

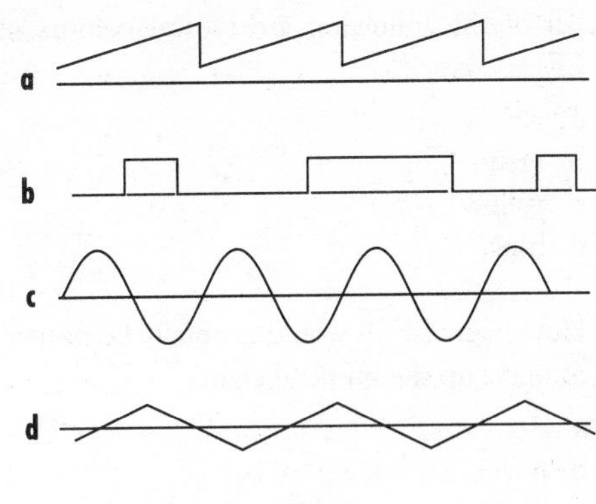

Fig. 7

21. In the UK the frequency of the ac supply is generated at 50 Hz. How many complete cycles occur in 0.2 s?

a 4
b 3
c 2
d 1.

22. What is the approximate **peak-to-peak** value for a sinusoidal waveform if its rms value is 240 V?

a 679 V
b 480 V
c 339 V
d 240 V.

23. Fig. 8 shows a circuit diagram of a light emitting diode (LED) which requires only 2 V at 10 mA to make it operate. What is the value of R if it is to operate safely on the 12 V supply?

a 1000 Ω
b 100 Ω
c 10 Ω
d 1 Ω.

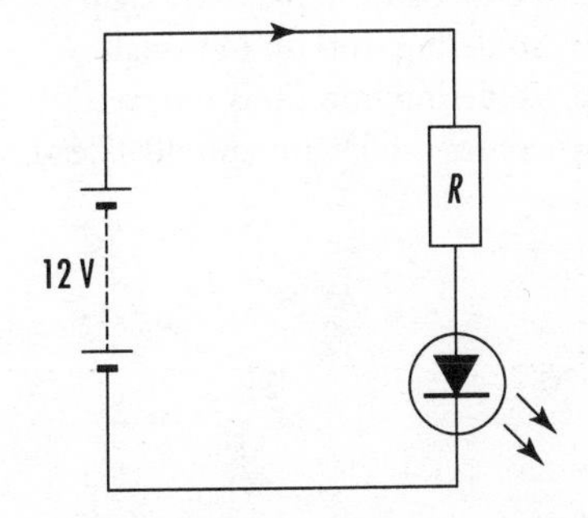

Fig. 8

24. All of the following are typical colours of LEDs *except*

a red
b green
c yellow
d blue.

25. How many LEDs would normally be needed to make up the numeral eight?

a 7
b 6
c 5
d 4.

26. Which one of the following is the correct order for carrying out soldering work?

1. wet the cleaning sponge
2. melt solder on the joint not the iron
3. make the mechanical joint before soldering
4. allow the soldering iron to heat up
5. make sure the sponge is kept wet.

a 5, 4, 3, 2, 1
b 1, 2, 3, 4, 5
c 4, 1, 3, 2, 5
d 3, 2, 1, 5, 4.

27. When soldering a sensitive electronic component with wire leads into a circuit, which one of the following precautions should you observe?

a joint should be soldered slowly
b thermal shunt should be used
c flux should be slightly corrosive
d heat-resistant gloves should be worn.

28. Which one of the following would be the most likely cause of a poor soldering joint?

a connection is made too tight
b soldering iron tip too small
c soldering iron tip is too wet
d solder is 60% tin and 40% lead.

29. Fig. 9 shows how a soldered joint is made. As soon as the solder runs into the joint the

a iron should be slightly twisted and blown cool
b solder and iron should be removed together
c iron should be removed before the solder
d solder should be removed before the iron.

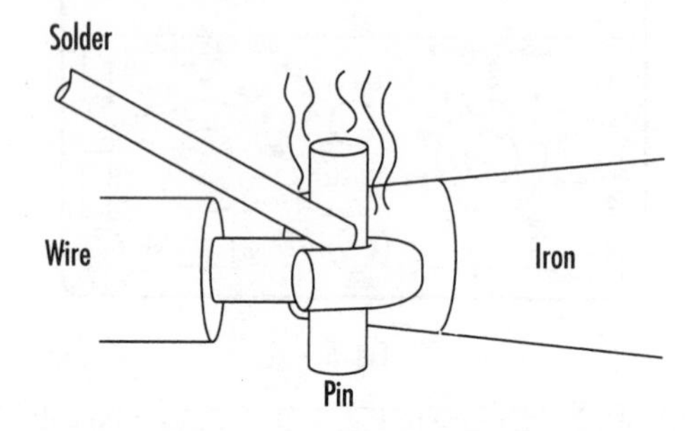

Fig. 9

30. In soldering work, all of the following are reasons why a poor joint may be formed, *except*

a the joint has not been properly cleaned
b insufficient flux has been used on the joint
c insufficient heat is supplied from the soldering iron
d the joint is not invisible under the solder.

31. Fig. 10 shows a method of testing a dry soldered joint. If confirmed, the voltmeter connection at A will give a reading of

a 12 V
b 6 V
c 3 V
d 0 V.

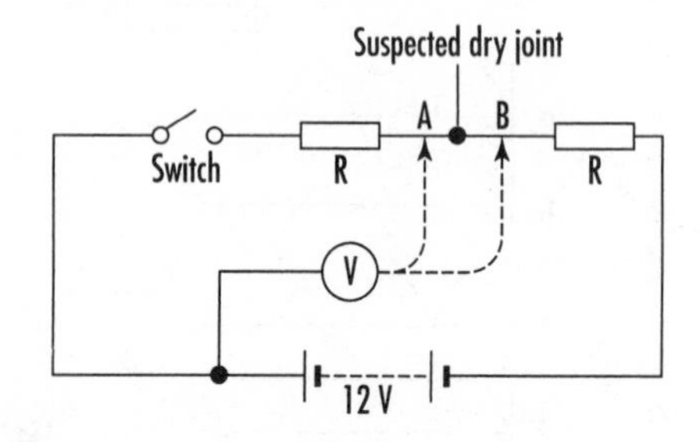

Fig. 10

32. With reference to Question 31 and Fig. 10, the voltmeter connection at B will give a reading of

a 12 V
b 6 V
c 3 V
d 0 V.

33. One of the advantages of making an **insulation displacement connection** over soldering and crimping methods is that the cable

a end does not require stripping
b can be flat or round in shape
c can be single or multicore
d cores are correctly numbered.

34. Fig. 11 shows a diagram of a double-wound transformer. The core is laminated to

a reduce power loss
b create more magnetism.
c avoid current saturation.
d avoid stray currents.

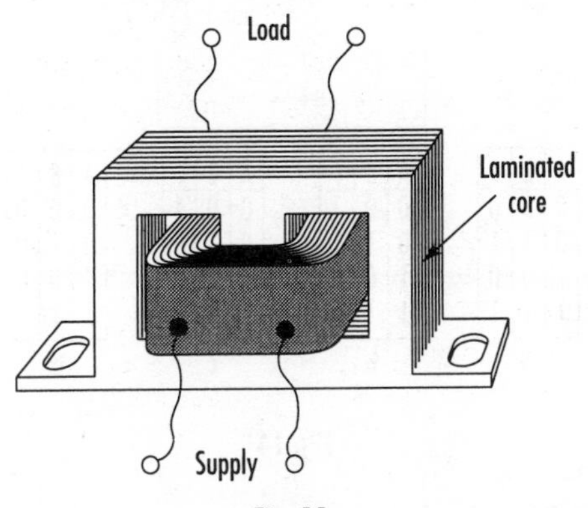

Fig. 11

35. An **air-cored inductor** is one that is chosen for its use in

a rectifying circuits
b radio tuning circuits
c discharge lamp circuits
d telephone circuits.

36. One of the advantages of using coaxial cable over other types of electronic cable is its

a low frequency use and use on dc supplies
b low resistance and high reactance property
c high frequency use and low electrical loss
d solid polythene insulation property.

37. In an amplifier, the amount by which a signal can be amplified is called the

a gain
b modulation
c transmission
d attenuation.

38. The ratio of the time that a rectangular waveform is high to the time that it is low is called the

a amplitude ratio
b energy loss ratio
c mark-to-space ratio
d signal ratio.

39. Which one of the following components possesses inductance?

a rectifier
b capacitor
c rheostat
d transformer.

40. The name of the semiconductor component abbreviated to FET is a

a triac
b unipolar transistor
c silicon thyristor
d thermistor.

41. When a pn junction diode is correctly wired in a dc circuit it acts like a

a rectifier
b one-way switch
c transducer
d regulator.

42. In electronics, a variable resistor is often called a

a potentiometer
b regulator
c stabilizer
d adjuster.

43. Which one of the following components has the ability to hold a charge?

a thermistor
b reactor
c capacitor
d inductor.

44. Fig. 12 shows the internal circuit plan of an IC package. What is the component symbol marked X?

- **a** rectifier
- **b** pot
- **c** silicon chip
- **d** amplifier.

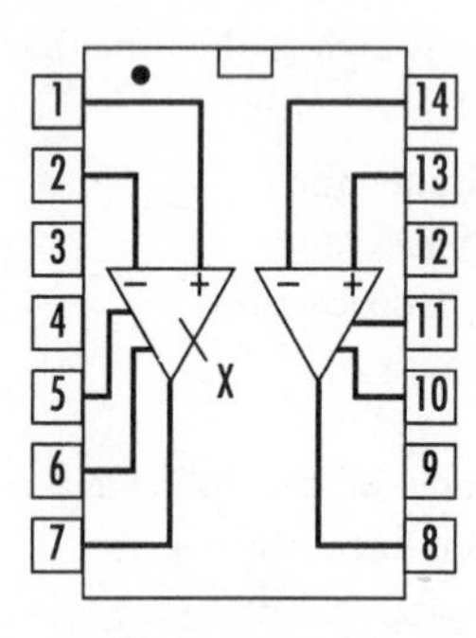

Fig. 12

45. When **harmonic waveforms** mix with the fundamental waveform they cause

- **a** attenuation
- **b** rectification
- **c** distortion
- **d** modulation.

46. Which one of the following is likely to be the formation of a square wave?

- **a** odd harmonics plus fundamental wave
- **b** even harmonics plus fundamental wave
- **c** two overlapping ac rectified waves
- **d** two overlapping complex waves.

47. What is the name of the numerical code that is used in logic gate circuits to indicate switching modes?

- **a** unitary code
- **b** binary code
- **c** analogue code
- **d** digital code.

48. Fig. 13 shows a typical digital signal waveform. The high point on the wave is called

- **a** logic 4
- **b** logic 2
- **c** logic 1
- **d** logic 0.

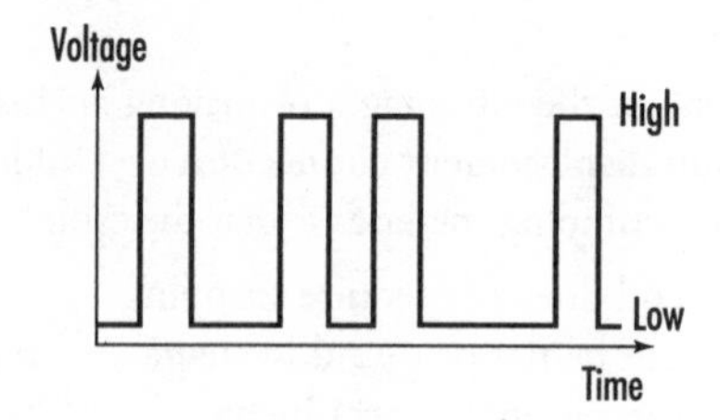

Fig. 13

49. Which one of the following truth tables in Fig. 14 represents an AND logic gate?

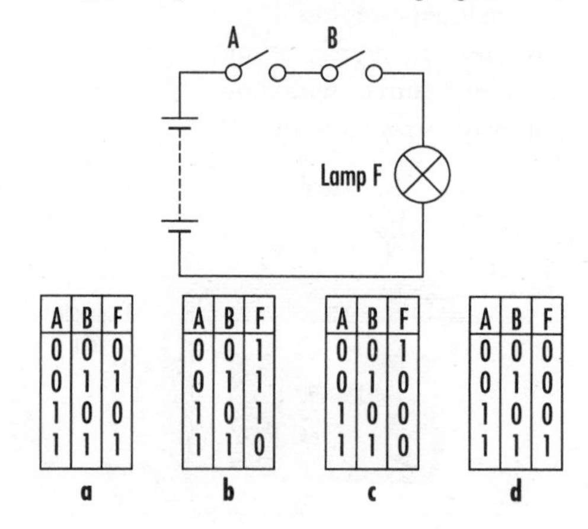

a

A	B	F
0	0	0
0	1	1
1	0	0
1	1	1

b

A	B	F
0	0	1
0	1	1
1	0	1
1	1	0

c

A	B	F
0	0	1
0	1	0
1	0	0
1	1	0

d

A	B	F
0	0	0
0	1	0
1	0	0
1	1	1

Fig. 14

50. The semiconductor component used to make the switching of a NOT logic gate is called a

- **a** transistor
- **b** triac
- **c** diac
- **d** zener diode.

ASSESSMENT 10.3

Answers, Hints and References

1. **c** It is not practical.
2. **b** *See* Ref. 27, p 25.
3. **c** *See* Ref. 27, p 28.
4. **b**.
5. **a** *See* Ref. 27, p 25.
6. **c** *See* Ref. 27, p 43.
7. **b** *See* Ref. 27, p 43.
8. **a** *See* Ref. 27, pp 4–5.
9. **a** *See* Ref. 27, p 24.
10. **d** *See* Ref. 27, p 87.
11. **c** *See* Ref. 27, p 96.
12. **d** *See* Ref. 6, pp 61–62.
13. **a** *See* Ref. 7, p 96.
14. **a** *See* Ref. 7, p 100.
15. **c** *See* Ref. 7, p 96.
16. **d** *See* Ref. 3, pp 107–109.
17. **c** *See* Ref. 3, pp 107–109.
18. **b** *See* Ref. 3, pp 107–109.
19. **c** *See* Ref. 7, p 99.
20. **b** *See* Ref. 6, p 66.
21. **d** *See* Ref. 6, pp 66–67.
22. **a** *See* Ref. 7, p 54.
23. **a** *See* Ref. 27, p 84.
24. **d** *See* Ref. 27, p 84.
25. **a** *See* Ref. 27, p 85.
26. **c** *See* Ref. 27, p 20.
27. **b** *See* Ref. 27, p 23, Fig. 2.12.
28. **b**.
29. **d** *See* Ref. 27, pp 20–21.
30. **d** *See* Ref. 27, p 21.
31. **d** *See* Ref. 27, p 21, Fig. 2.7.
32. **a** *See* Ref. 27, p 27.
33. **a** *See* Ref. 27, p 28.
34. **a** *See* Ref. 7, p 91.
35. **b** *See* Ref. 6, p 58.
36. **c** *See* manufacturers' literature.
37. **a** *See* Ref. 27, p 186.
38. **c** *See* Ref. 6, p 68.
39. **d** *See* Ref. 7, p 89.
40. **b** *See* Ref. 6, pp 62–63.
41. **b** *See* Ref. 6, p 60.
42. **a** *See* Ref. 6, p 56.
43. **c** *See* Ref. 6, p 56.
44. **d** *See* Ref. 6, p 13, Fig. 1.9.
45. **c** *See* Ref. 6, p 68.
46. **a** *See* Ref. 6, p 68.
47. **b** *See* Ref. 27, p 115.
48. **c** *See* Ref. 27, p 115.
49. **d** *See* Ref. 27, p 117.
50. **a** *See* Ref. 27, p 86.

REFERENCES

1. Lewis M. (1995) *Electrical Installation Competences. Part 1 Studies: Theory.* Cheltenham: Stanley Thornes (Publishers) Ltd.
2. Lewis M. (1995) *Electrical Installation Competences. Part 2 Studies: Theory.* Cheltenham: Stanley Thornes (Publishers) Ltd.
3. Lewis M. (1994) *Electrical Installation Technology 3: Advanced Work.* Cheltenham: Stanley Thornes (Publishers) Ltd.
4. *IEE Wiring Regulations – Regulations for Electrical Installations (BS7671) 1992.* Institution of Electrical Engineers.
5. *UK Electricity 1994.* Electricity Association Services Ltd.
6. Lewis M. (1994) *Electrical Installation Competences. Part 1 Studies: Science.* Cheltenham: Stanley Thornes (Publishers) Ltd.
7. Lewis M. (1995) *Electrical Installation Technology. Part 2: Science and Calculations.* Cheltenham: Stanley Thornes (Publishers) Ltd.
8. Lewis M. (1993) *Electrical Installation Competences. Part 2 Studies: Science.* Cheltenham: Stanley Thornes (Publishers) Ltd.
9. *Health and Safety (First Aid) Regulations 1981.*
10. *Electricity at Work Regulations 1989.* Health and Safety Executive Statutory Regulations.
11. *Health and Safety (Safety Signs and Signals) Regulations 1996.* EU Directive 92/58/EEC.
12. Lewis M. (1992) *Electrical Installation Technology. Part 1: Theory and Regulations.* Cheltenham: Stanley Thornes (Publishers) Ltd.
13. *INDG 4 First Aid at Work.* Health and Safety Executive leaflet.
14. *INDG Emergency Aid.* Health and Safety Executive leaflet.
15. *Control of Substances Hazardous to Health Regulations 1988.* Health and Safety Executive Regulations.
16. *INDG 93 Solvents and You.* Health and Safety Executive leaflet.
17. *Health and Safety at Work Act 1974.* Health and Safety Executive Regulations.
18. *IEE On-Site Guide. (1995)* Institution of Electrical Engineers.
19. Lewis M. (1991) *Electrical Installation Competences. Part 1 Studies: Practical.* Cheltenham: Stanley Thornes (Publishers) Ltd.
20. Lewis M. (1994) *Electrical Installation Competences. Part 2 Studies: Practical.* Cheltenham: Stanley Thornes (Publishers) Ltd.
21. *JT Ltd Electrical Installation, Study Notes books 1, 2, 3 and 4.* Construction Industry Training Board.
22. *Noise at Work Regulations.* (1989) Health and Safety Executive Regulations.
23. *Office, Shops and Railways Premises Act 1963.* Health and Safety Executive Regulations.
24. *Manual Handling Operations Regulations,* Guidance on Regulations 1992 (L23).
25. *INDG 109/110 Lighten the Load.* Health and Safety Executive leaflet.
26. National Inspection Council for Electrical Installation Contracting *Snags and Solutions.* (1992) London: Reed Business Publishing.
27. Linsley T (1992) *Electronics for Electricians.* London: Edward Arnold.
28. *IEE Guidance Notes, No. 3, Inspection and Testing.* (1992) Institution of Electrical Engineers.
29. *IEE Guidance Notes, No. 1, Selection and Erection of Equipment.* (1993) Institution of Electrical Engineers.